IEE Electromagnetic Waves Series 18

Series Editors: Professor P.J.B. Clarricoats
E.D.R. Shearman and J.R. Wait

Corrugated horns for microwave antennas

Previous volumes in this series

Volume 1	Geometrical theory of diffraction for electromagnetic waves Graeme L. James
Volume 2	Electromagnetic waves and curved structures Leonard Lewin, David C. Chang and Edward F. Kuester
Volume 3	Microwave homodyne systems Ray J. King
Volume 4	Radio direction-finding P.J.D. Gething
Volume 5	ELF communications antennas Michael L. Burrows
Volume 6	Waveguide tapers, transitions and couplers F. Sporleder and H.G. Unger
Volume 7	Reflector antenna analysis and design P.J. Wood
Volume 8	Effects of the troposphere on radio communications Martin P.M. Hall
Volume 9	Schuman resonances in the earth-ionosphere cavity P.V. Bliokh, A.P. Nikolaenko and Y.F. Filippov
Volume 10	Aperture antennas and diffraction theory E.V. Jull
Volume 11	Adaptive array principles J.E. Hudson
Volume 12	Microstrip antenna theory and design J.R. James, P.S. Hall and C. Wood
Volume 13	Energy in Electromagnetism H.G. Booker
Volume 14	Leaky feeders and subsurface radio communications P. Delogne
Volume 15	The Handbook of Antenna Design Volume 1 Editors; A.W. Rudge, K. Milne, A.D. Olver, P. Knight
Volume 16	The Handbook of Antenna Design Volume 2 Editors: A.W. Rudge, K. Milne, A.D. Olver, P. Knight
Volume 17	Surveillance Radar Performance Prediction P. Rohan

Corrugated horns for microwave antennas

P.J.B. Clarricoats and A.D. Olver

Peter Peregrinus Ltd
On behalf of The Institution of Electrical Engineers

Published by: Peter Peregrinus Ltd., London, UK.

Clarricoats, P. J. B.
Corrugated horns for microwave antennas.
—(IEE electromagnetic waves series; 18)
1. Microwave antennas 2. Antennas, Reflector
3. Antenna feeds
I. Title II. Olver, A. D. III. Series
621.381'33 TK7871.6

ISBN 0 86341 003 0

Printed in England by Short Run Press Ltd., Exeter

Contents

Preface viii

Acknowledgments ix

1 Introduction 1
- 1.1 Historical observations 1
- 1.2 Organisation of the book 3

2 Introduction to hybrid-mode feeds 5
- 2.1 Illumination of a reflector antenna 5
- 2.2 Focal field of a reflector 10

3 Propagation and radiation characteristics of cylindrical corrugated waveguides 20
- 3.1 Introduction 20
- 3.2 Propagation in cylindrical corrugated waveguide 22
- 3.3 Fields in cylindrical corrugated waveguide 35
- 3.4 Attenuation in corrugated waveguide 40
- 3.5 Radiation from corrugated waveguides 43
  - 3.5.1 Introduction 43
  - 3.5.2 Radiation from a circular aperture 45
- 3.6 Junctions in corrugated waveguides 50
- 3.7 Multimode corrugated waveguides 54

4 Propagation and radiation characteristics of conical corrugated waveguides 58
- 4.1 Introduction 58
- 4.2 Modal characteristics of corrugated conical horns 58
- 4.3 Radiation from corrugated conical horns by the spherical wave expansion method 65
  - 4.3.1 The spherical wave expansion method 65
  - 4.3.2 Frequency dependence of crosspolar radiation field 68

4.4 Generation of higher order modes 69
4.4.1 Scattering at an abrupt discontinuity 71
4.4.2 Higher order mode generation at the throat 73
4.4.3 Mode conversion along a horn with constant flare angle 73
4.4.4 Mode conversion along a horn with variable flare angle 78
4.5 Radiation from corrugated horns by the Kirchhoff–Huygen method 86
4.6 Comparison of predicted and measured radiation patterns 90
4.7 Radiation from corrugated horns by cylindrical waveguide approximation 90
4.8 Radiation from conical corrugated horns by means of the Laguerre–Gaussian expansion method 95

5 Design of cylindrical and conical corrugated horns 97
5.1 Introduction 97
5.2 Copolar radiation characteristics 100
5.3 Crosspolar radiation characteristics 114
5.4 Flare section of horn 122
5.5 Throat region of horn 128
5.6 Gain and directivity 135
5.7 Efficiency when used as a feed for a reflector 137
5.8 Horn design 141
5.8.1 Large aperture, narrow flare angle horns 141
5.8.2 Small aperture horns 143
5.8.3 Wide flare angle horns 145
5.8.4 Multimode horns 150
5.8.5 Broadband horns 153
5.8.6 Multifrequency horns 154
5.8.7 Compact horns 155
5.9 Numerical prediction of performance 156
5.9.1 Propagation characteristics 157
5.9.2 Radiation characteristics 160

6 Manufacture and testing of corrugated horns 163
6.1 Manufacture of corrugated horns 163
6.2 Testing of corrugated horns 169
6.2.1 Introduction 169
6.2.2 Types of test range 171
6.2.3 Anechoic chamber design for low crosspolarisation horn measurement 172
6.2.4 Assessment of test range performance 175
6.2.5 Measurement of copolar and crosspolar patterns 177
6.2.6 Measurement of phase patterns and gain 180

7 Rectangular and elliptical corrugated horns 181
7.1 Introduction 181
7.2 Rectangular corrugated horns 181
7.2.1 Background 181
7.2.2 Two-walled rectangular corrugated waveguide 183
7.2.3 Four-wall rectangular corrugated waveguide 188
7.2.4 Radiation characteristics of rectangular corrugated horns 190
7.3 Elliptical corrugated horns 193

Appendix 199

Bibliography 213

Programs 222

Index 227

Preface

The idea for a text on the theory and design of corrugated horns became apparent to us during the late 1970s, as this type of feed became more and more widely used in antennas. Generally, the corrugated horn offers the reflector antenna designer the potential for producing antennas with higher efficiency, lower cross-polarisation and lower sidelobes, and it can be fairly said to lie at the heart of most optimally designed reflector antennas.

The book contains material which should be of value both to the designer and the research worker. Parts have been presented previously by us, at various vacation schools and in invited papers at conferences, over the last ten years, but the text also contains many new results. Some of these are drawn from the theses of our research students and from the work of others in our Electromagnetics Applications Group at Queen Mary College in the University of London. To these people we offer our warmest thanks; they are separately identified in the acknowledgments which follows. A description of the organisation of the book is to be found in Chapter 1, alongside a brief history of the subject.

Acknowledgments

The authors are indebted to many who have helped in the production of this text. They would especially like to thank the following members of the Queen Mary College Electromagnetics Applications Group, whose theoretical and experimental contributions over the period 1968–83, have proved so valuable: Drs. Al-Hariri, Chan, Chong, Elliott, Hockham, Mahmoud, Parini, Poulton, Saha, Salema and Seng. They also thank Mr. Kolb who undertook many measurements in our Antenna Laboratory; Messrs Ede, Fairbrass and Goose, whose excellent craftsmanship produced many intricate feeds, and Miss Sandra O'Callaghan, who typed the manuscript with great care and attention to detail. One of us (PJBC) is grateful for the encouragement of his wife Phyl during the preparation of the text.

Chapter 1

Introduction

1.1 Historical observations

There are two main reasons for the existence of corrugated horns as feeds for reflector antennas. First, they exhibit radiation pattern symmetry, which offers the potential for producing antennas with high gain and low spillover; secondly, they radiate with very low crosspolarisation, which is essential in dual-polarisation systems. The former property provided the motivation for Kay [88, 89] who conceived the wide-angle corrugated horn in 1962 while working at TRG in the USA. It is said that he was studying the effect of quarter-wavelength chokes at the horn aperture and found that, by adding more than one, the pattern symmetry improved. He noted invariance of the chokes to field orientation and thus coined the term scalar feed to describe his horn. At almost the same time, and quite independently, Minnett and Thomas [101], at CSIRO in Australia, pursued the radiation properties of corrugated waveguide feeds for use in radiotelescopes. They were motivated by efficiency and polarisation considerations. In this latter respect, Rumsey [134] contributed during a stay at CSIRO in 1966, by observing general properties of hybrid modes in respect of polarisation purity.

Slightly later another but related, hybrid-feed, the dielectric cone or dielguide, was devised by Barlett and Mosely [179, 180] at Radiation Incorporated in USA. This feed, which has some properties similar to optical waveguides, radiates after the manner of corrugated horns.

It is said that Barlett and Mosely's discovery was made while experimenting with a fibre glass tube in order to support a sub-reflector in a Cassegrain antenna. Apparently, the introduction of expanded foam into the tube was found to improve the feed performance. A qualitative description of the operation was provided in their patent applications which followed. A quantitative description came from Clarricoats and Salema [182], who made a comprehensive study in the early seventies. The authors' group at Queen Mary College was to the fore in analysing the corrugated horn and waveguide, beginning their investigations in 1968 and continuing them to the present day. They were first to publish a theory for the wide-angle corrugated horn, although Viggh [154] at TRG had made a contemporary study which appeared as an internal report in 1969. Minnett and his colleagues [102, 145] at CSIRO were also working on the wide-angle horn at about the same time as was Jeuken and his colleagues [80, 82, 85, 86], at the Technical University of Eindhoven.

In India, Narasimhan [107, 121] produced useful approximate methods of design suitable for those with limited computer resources. It appears that Bryant [19, 20] was the first to publish an account of rectangular corrugated horns, describing work undertaken by him at the Plessey Company in England, in the late sixties. Peters and his colleagues [57, 58, 92, 100, 143] at Ohio State University analysed relevant corrugated surfaces but in general the rectangular corrugated horn is not susceptible to closed-form analysis in the way that horns of circular cross-section can be treated. Thus, there is a great paucity in the literature of descriptions of feeds of this class and all early designs were empirical. Bryant commented in his paper that corrugated horns might have high attenuation, and this remark prompted Clarricoats and Saha [29] to examine the attenuation properties of circular corrugated waveguides. Contrary to Bryant's suggestion, these waveguides exhibit, for the lowest order hybrid HE_{11} mode, an attenuation even lower than that of the TE_{01} mode in a smooth-wall waveguide of comparable size. During the period 1970–75, a very detailed theoretical and experimental study of the attenuation properties of corrugated waveguides was conducted by the present authors and their reserach students. Although, all the basic properties were confirmed, it proved very difficult to manufacture corrugated waveguides in long lengths and the project was terminated around 1976. However, the very detailed investigation which included full space-harmonics analysis of the periodic waveguide, provided an excellent base for subsequent investigations of both the propagating and radiation properties of corrugated horns used as feeds.

By 1975, interest in dual-polarised feeds was growing under the influence of the satellite communication market. This provided a major impetus for further work on corrugated feeds. Tracking requirements also prompted work on dual-mode feeds and mention should be made in this selective history of the substantial contribution from workers at ERA Technology Limited. Two important feed designs were developed within their group; the matched-feed, for use with offset reflectors, conceived by Rudge and Adatia [189, 190], and the dual-depth corrugated horn investigated by Ghosh [63, 64].

Around 1975, satellite broadcasting led to shaped beam requirements which stimulated research on elliptical corrugated waveguides. There the work of Jeuken and Vokurka [78, 83] is noteworthy. However, the difficulties of analysis and manufacture have limited application of this work.

We are now close to the present day, and many names have been missed from this brief account, although their valuable contributions will be identified as the reader delves further into our text. However, before concluding this section we should comment on the numerous contributions from Thomas, James and their colleagues [74–77, 101, 102, 144–152], at CSIRO, whose interest in the subject of corrugated horns has traversed the same period as the present authors, and whose work has been widely reported in the literature. Also Dragone [53–56] at Bell Telephone Laboratories, whose 1977 papers provided substantial added insight.

1.2 Organisation of the book

Chapter 2 lays the foundation for the chapters which follow, by exploring the desired properties of reflector antennas, both from the standpoint of radiation patterns and focal fields. These studies reveal the optimum qualities of hybrid mode feeds of which the corrugated horn provides the most important example. Chapters 3 and 4 present in some detail the theory underlying the propagation and radiation characteristics of, respectively, cylindrical and conical corrugated horns. The former is amenable to exact analysis although in many cases it is sufficient to use the surface-impedance approximations to describe the boundary conditions at the corrugated waveguide wall. By contrast, there is no exact formulation for the conical corrugated horn, but, notwithstanding, very accurate prediction methods have been developed using, in particular, a spherical mode treatment which extends to hybrid modes, the pure-mode descriptions found for example, in the text by Harrington [183]. Here the authors have drawn heavily on the work of their colleague Mahmoud [95].

Chapter 5 leads us in to design. It builds on the work of Chapters 2, 3 and 4, but, if desired, it can be read almost independently of them. Mathematics appears sparingly so this chapter should appeal strongly to the person who has to build horns and does not have the time to cross all the theoretical bridges of earlier chapters. Chapter 5 draws heavily on experimental results and computer programs from QMC (two of which are included). The authors recognise that many others will have obtained similar results. There are obvious advantages to be gained in a text when results are drawn mainly from one source.

Chapter 6 addresses the problem of horn manufacture. It is to be hoped that our observations will prove helpful but we shall be glad to learn from readers of any techniques which they have found satisfactory and which we have not reported.

Chapter 6 also explores methods for the measurement of corrugated horns. As stated there, these are essentially the methods for the precision measurement of any small antenna except that the corrugated horn does exhibit some rather special properties. As these properties are what the customer pays for, e.g. low cross-polarisation, low sidelobes and pattern symmetry, it is understandable that he will want to know rather precisely how well his horn performs. Chapter 6 provides answers which will also serve the interests of those seeking a general introduction to microwave antenna measurements. More information is to be found, for example, in *The Handbook of Antenna Design.*

Chapter 7 explores the characteristics of corrugated horns of rectangular and elliptical cross-section. In contrast with the circular case the non-circular corrugated waveguide is not amenable to exact analysis. However, approximate methods have been tried with limited success and the chapter presents a summary with pertinent results. These feeds are difficult to manufacture and their performance is generally inferior to their circular counterpart. They are used only in those applications where a primary feed pattern is needed with differing beamwidths in the principal planes.

The book has an appendix in which the space-harmonic analysis of the circular corrugated waveguide is presented in detail.

Before concluding this introduction we must comment on certain feeds which have not been included in our text. First, the Potter horn [188] in which a hybrid field is synthesised from pure modes in a multimode circular waveguide. The prototype employed two modes, the TE_{11} and TM_{11}, but subsequent versions were considered in which additional modes were employed. By the nature of its operation the structure is narrow band but it has the merit of relative simplicity of construction. It is sometimes used as an alternative to corrugated horns where bandwidth and beamwidth considerations allow. An account of the Potter horn is also to be found in papers collected in the reprint volume edited by Love [94].

Another group of feeds which are not considered are those in which the corrugations form chokes exterior to the primary pure-mode circular waveguide. Feeds of this kind have been described by Koch [91], Wohlleben, Mattes and Lochner [164], and also Scheffer [136]. A prototype structure, the 90° flare-angle corrugated horn, was analysed by Hockham [71], but he considered only the case where the plane of the primary waveguide coincided with that of the corrugations. Some features present in the other feeds were apparent but not all the details which made them of practical importance as primary feeds for paraboloid antennas. Unfortunately, the general case in which the chokes do not lie in the plane of the primary waveguide represents a most difficult electromagnetic problem that awaits solution. These structures are only related to the hybrid feeds discussed in this text because they excite spherical hybrid modes in the vicinity of the radiator. The feeds described herein all support hybrid modes as propagating fields before radiating at their apertures.

Finally, our text ends with a bibliography which attempts to be comprehensive at the time of writing.

Chapter 2

Introduction to hybrid-mode feeds

2.1 Illumination of a reflector antenna

The most important use in antennas for a hybrid-mode waveguide or horn is as a feed for a reflector, so we begin this chapter by considering, in general terms, how the feed influences antenna performance. Fig. 2.1 shows a photograph of a horn feed in a Cassegrain antenna forming part of a satellite-communication earth-station while Fig. 2.2 shows a dielectric-cone feed in an antenna for a similar application. Although different in appearance and construction, the corrugated horn and dielectric-cone feed support nearly identical hybrid fields and in this chapter we shall use the term hybrid-mode feed to cover both types.

The antennas shown in Figs. 2.1 and 2.2 produce pencil beams, and to be effective, their gain should be as high as possible, consistent with an acceptable sidelobe envelope. Furthermore, if, as is now usual, the antenna is required to transmit and receive orthogonally polarised waves, there will also be an exacting specification on the crosspolar radiation pattern. The antenna gain in the boresight direction G, is related to the aperture area A and wavelength through

$$G = \frac{4\pi A}{\lambda^2}\eta \tag{2.1}$$

where η, the antenna efficiency, is the product of a number of terms identified in Table 2.1. The table shows how the first four terms depend on the feed radiation pattern. If the feed radiation pattern is symmetric, i.e. the E-plane and H-plane patterns are identical, $\eta_x = 1$ and furthermore, if the feed has a common phase centre for both planes and is properly located, $\eta_p = 1$. This leaves the illumination and spillover losses as the principal causes of gain degradation due to the feed and even in an ideal antenna there must be a compromise between these last two factors. If the illumination is uniform $\eta_I = 1$ but then the spillover efficiency η_s, for any realisable feed, would be too low. Not only would this reduce the antenna gain but the antenna sidelobe levels and antenna noise-temperature would be too high giving rise to problems both in earth-station and microwave-relay applications. In passing we note that by using a dual-shaped reflector design, it is possible to

Fig. 2.1 *Intelsat type B earth-station antenna with a narrow flare-angle corrugated horn (Courtesy Andrew Antennas)*

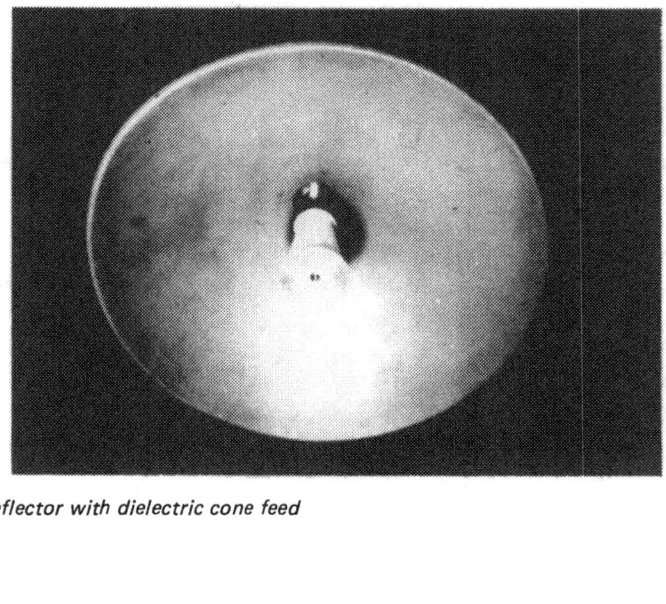

Fig. 2.2 *Reflector with dielectric cone feed*

achieve the best possible compromise between these two factors since by controlling the subreflector shape, the illumination can be made nearly uniform over most of the main reflector while being heavily tapered near to the reflector edge.

We can see from the above that an *ideal* feed for a symmetric reflector antenna is one which has a symmetric radiation pattern and this also ensures zero cross-polarisation. Many different reflector geometries exist and since systems operate over quite wide bandwidths, sometimes in two or more frequency bands, we require our feeds also to have wide bandwidths, as well as a variety of beamwidths, corresponding to different reflector f/D ratios. Only the *hybrid-mode feed* offers a means to approach all these objectives simultaneously.

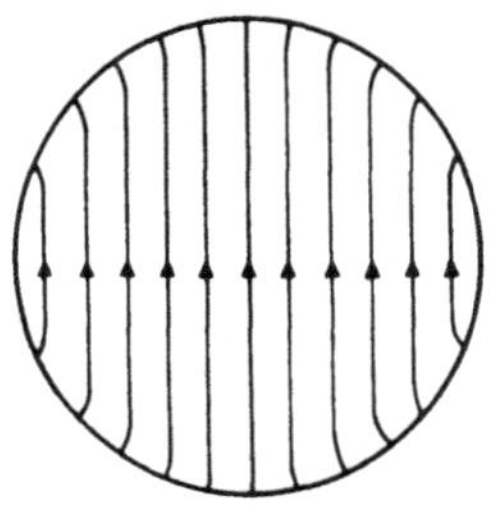

Fig. 2.3 *Transverse field patterns for dominant HE_{11} mode in cylindrical waveguide of radius r_1*

To emphasise this statement let us examine, for the dominant mode, the electric field in the aperture of a hybrid-mode waveguide, such as that of Fig. 2.3. The derivation of the fields will come later. The representation is appropriate to a corrugated waveguide of internal radius r_1. It is also a good approximation for a dielectric waveguide of radius r_1, if $kr_1 \gg 1$ (aperture large compared to a wave-length) for which the field is then mainly confined to the dielectric region $r < r_1$.

$$\mathbf{E} = AJ_0(Kr)\hat{i}_x - \left(\frac{X-Y}{4}\right)\frac{U_0^2}{kr_1} J_2(Kr)(\cos 2\phi\hat{i}_x + \sin 2\phi\hat{i}_y) \tag{2.2}$$

$J_n(Kr)$, $n = 0, 2$ is a Bessel function of the first kind and order n, $\hat{i}_{x,y}$ are unit vectors in the x and y directions, U_0 is a normalised transverse wavenumber, K and k are transverse and free-space wavenumbers, respectively, X and Y are the normalised reactance and admittance of the boundary at $r = r_1$. In Fig. 2.3, the boundary is that of a circular cylinder so,

$$X = -j\frac{Z_\phi}{Z_0} = -j\frac{E_\phi}{H_z}\left[\frac{\epsilon_0}{\mu_0}\right]^{1/2} \tag{2.3}$$

$$Y = -j\frac{Z_0}{Z_z} = +j\frac{H_\phi}{E_z}\left[\frac{\mu_0}{\epsilon_0}\right]^{1/2} \tag{2.4}$$

and the fields in eqns. 2.3 and 2.4 are to be evaluated at the boundary at $r = r_1$.

Table 2.1 *Components of efficiency of reflector antennas*

Illumination efficiency

$$\eta_I = 2\cot^2\frac{\theta_0}{2}\,\frac{|\int_0^{\theta_0}(|F_E| + |F_H|)\tan(\theta/2)\mathrm{d}\theta|^2}{\int_0^{\theta_0}(|F_E| + |F_H|^2)\sin\theta\,\mathrm{d}\theta}$$

Spillover efficiency

$$\eta_s = \frac{\int_0^{\theta_0}(|F_E|^2 + |F_H|^2)\sin\theta\,\mathrm{d}\theta}{\int_0^{\pi}(|F_E|^2 + |F_H|^2)\sin\theta\,\mathrm{d}\theta}$$

Phase error efficiency

$$\eta_p = \frac{|\int_0^{\theta_0}(F_E + F_H)\tan(\theta/2)\,\mathrm{d}\theta|^2}{|\int_0^{\theta_0}(|F_E| + |F_H|)\tan(\theta/2)\,\mathrm{d}\theta|^2}$$

Crosspolarisation efficiency

$$\eta_x = \frac{\int_0^{\theta_0}(|F_E| + |F_H|)^2\sin\theta\,\mathrm{d}\theta}{2\int_0^{\theta_0}(|F_E|^2 + |F_H|^2)\sin\theta\,\mathrm{d}\theta}$$

Blockage efficiency

$$\eta_B = \frac{\int_{\theta_B}^{\theta_0}(|F_E| + |F_H|)\tan(\theta/2)\,\mathrm{d}\theta|^2}{\int_0^{\theta_0}(|F_E| + |F_H|)\tan(\theta/2)\,\mathrm{d}\theta|^2}$$

Surface error efficiency

$$\eta_E = \exp\,[-(4\pi\epsilon/\lambda)^2] \qquad \epsilon = \text{RMS surface error}$$

Total efficiency

$$\eta = \eta_I\eta_s\eta_p\eta_x\eta_B\eta_E$$

In the above, the feed radiation pattern is represented by

$$E = \frac{e^{-jkp}}{P}(F_E\sin\phi + F_H\cos\phi)$$

F_E and F_H are the E plane and H plane amplitude patterns, respectively. P is the distance from the feed to the reflector surface.

From eqn. 2.2 we see that if the term $(X - Y)$ vanishes, the aperture field is independent of the angular variable ϕ and is also free of crosspolarisation. Furthermore, because the radiated field is the Fourier Transform of the aperture field, these properties are transferred to the radiation field. The condition $(X - Y) = 0$ can be realised either with X and Y finite and equal or with X and Y both zero. We now examine the condition in two types of hybrid-mode waveguide, noting in passing that the condition is never satisfied in a conventional pure-mode waveguide.

In a corrugated waveguide X is generally zero, or nearly so, and

$$Y = \tan\left(m\frac{\pi}{2}\frac{\Delta f}{f_0}\right) \tag{2.5}$$

where Δf measures the frequency deviation from the resonant value f_0, at which Y vanishes when m is an odd integer. Thus a corrugated waveguide does satisfy the constraint required of an ideal hybrid-mode feed but only at certain frequencies. Usually the corrugation depth is made approximately $\lambda/4$ at the centre of the operating band, then $m = 1$ in the expression for Y. It is an unfortunate property of corrugated waveguides that the condition for pattern symmetry and zero cross-polarisation occurs only at specific frequencies, but eqn. 2.2 does show that the relative effect of the ϕ independent terms, which are non-zero away from resonance, decrease as $1/kr_1$. This is a general property of cylindrical hybrid-mode feeds for which it can be stated that one with a large aperture gives an inherently better performance than a feed of small aperture, but it will of course produce a narrow beamwidth pattern unsuitable for prime-focus applications.

For a dielectric waveguide, if $kr_1 \gg 1$ and the refractive index differs by only a small amount Δn from unity, X and Y are given by:

$$X = \frac{1}{(2\Delta n)^{1/2}} \tag{2.6}$$

$$Y = \frac{1}{(2\Delta n)^{1/2}}(1 + 2\Delta n) \tag{2.7}$$

Thus,

$$(X - Y) = -(2\Delta n)^{1/2} \tag{2.8}$$

and we see that the optimum condition is approached when Δn is made very small. The condition is also frequency independent if, as is reasonable, the dielectric is free from dispersion over the operating frequency band. However, unless Δn is sufficiently large it is impossible to efficiently launch the dominant mode of the dielectric waveguide and a significant amount of energy will escape directly from the launcher. In practice a compromise must be reached. As an example, with $\Delta n/n = 0{\cdot}025$, the performance of the two types of hybrid-mode feed become equal when $(\Delta f/f_0)$ is 13%. Nearer to resonance the corrugated waveguide is superior. Other things being equal, the designer can choose, for example, between a lower crosspolarisation over a narrower band with a corrugated feed, or a somewhat higher crosspolarisation over a much wider band with the dielectric feed.

Some radar- and most satellite-broadcast antennas have non-circular cross-section in order that a shaped beam can be generated. A feed with a non-circular cross-section is required to generate an asymmetric pattern on transmit, or match an asymmetric focal field on receive. In spite of the asymmetry, orthogonal states of polarisation can exist and are frequently demanded with comparable specifications as for spot beams.

Both rectangular and elliptical cross-section corrugated horns have been used as feeds and rectangular cross-section dielectric waveguide feeds have also been studied. These structures can all be made to exhibit linear polarisation for the dominant mode. This property has been shown by Dragone [55] to be quite generally approached as the waveguide aperture dimensions become very large, provided there exists an axis of symmetry in the transverse plane.

2.2 Focal field of a reflector

The desirable features of hybrid-mode waveguides as feeds can be illustrated alternatively by examining the focal fields of a reflector. When a plane wave illuminates the reflector, as under normal receiving conditions, the field in the focal region can be predicted by means of physical optics. Fig. 2.4 shows the reflector configuration.

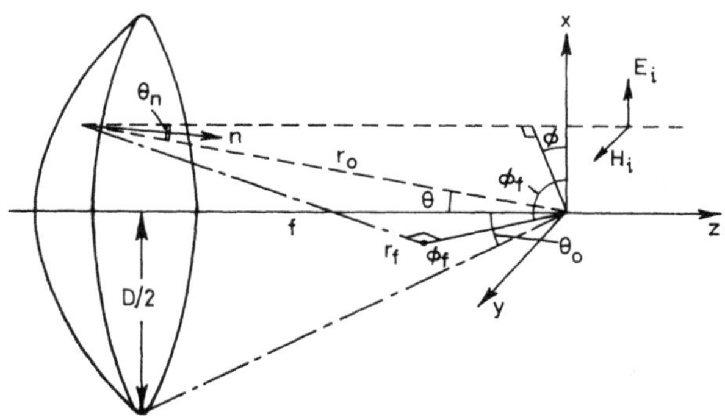

Fig. 2.4 *Reflector configuration*

The incident plane wave is assumed to have its electric field linearly polarised in the x direction, then the focal region electric field obtained by integrating the surface current **K** over the reflector surface, is

$$\mathbf{E} = \frac{j\omega\mu_0}{4\pi}\int_S [\mathbf{K} - (\mathbf{K}\hat{i}_r)\hat{i}_r]\,\frac{\epsilon^{-jkr}}{r}\,dS \tag{2.9}$$

Here, in the physical optics approximation,

$$\mathbf{K} = 2\hat{n} \times \mathbf{H}_i \tag{2.10}$$

and

$$H_i = -H_0 \hat{y} \epsilon^{-jkr_0 \cos\theta} \tag{2.11}$$

H_i is the incident magnetic field of the plane wave and the integration is over the reflector surface. Following Minnett and Thomas [186], it is possible to show that the focal-region transverse electric and magnetic fields at $z = 0$ have the form

$$\mathbf{E} = j\omega\mu_0 H_0 \{\hat{i}_x I_1 + I_2(\cos 2\phi_f \hat{i}_x + \sin 2\phi_f \hat{i}_y)\} \tag{2.12}$$

$$\mathbf{H} = j\omega\mu_0 \frac{H_0}{Z_0} \{\hat{i}_y I_1 + I_2(\sin 2\phi_f \hat{i}_x - \cos 2\phi_f \hat{i}_y)\} \tag{2.13}$$

The quantities I_1 and I_2, which are, respectively, proportional to the copolar and crosspolar fields, are given by

$$I_1 = \int_0^{\theta_0} \sin\theta J_0(kr_f \sin\theta)d\theta \, [f\epsilon^{-j2kf}] \tag{2.14}$$

$$I_2 = \int_0^{\theta_0} \sin\theta \tan^2\frac{\theta}{2} J_2(kr_f \sin\theta)d\theta \; [f\epsilon^{-j2kf}] \tag{2.15}$$

Later we shall show that eqn. 2.2 for the transverse electric field in a hybrid-mode waveguide can be recast in the form

$$\mathbf{E} = A\{\bar{I}_1 \hat{i}_x + \bar{I}_2(\cos 2\phi \hat{i}_x + \sin 2\phi \hat{i}_y)\} \tag{2.16}$$

and we can thus identify $\bar{I}_1$ and $\bar{I}_2$ as the kernels of the integrals I_1 and I_2. The field in the focal region is now recognisable as an angular spectrum of cylindrical hybrid waves. We gain insight into the meaning of this statement when we examine the form of the transverse-electric field in the focal region of a paraboloidal reflector possessing a representative f/D ratio as shown in Fig. 2.5. There the direction of the field is indicated while Figs. 2.6(a) and (b) show for various values of f/D the copolar and crosspolar intensity of the field. For a long focal-length reflector (θ_0 small), the field is almost entirely linearly polarised everywhere although the direction of the field alternates as the radial distance from the focus r_f is increased. For this case $I_2 \cong 0$ and the focal field has the form

$$E_x = 2E_0 \left(\frac{\pi D^2}{4f\lambda}\right) \frac{J_1(kr_f\theta_0)}{kr_f\theta_0} \tag{2.17}$$

The above equation corresponds to the scalar solution obtained by the 19th-century physicist George Airy, for the distribution of light in the focal region of

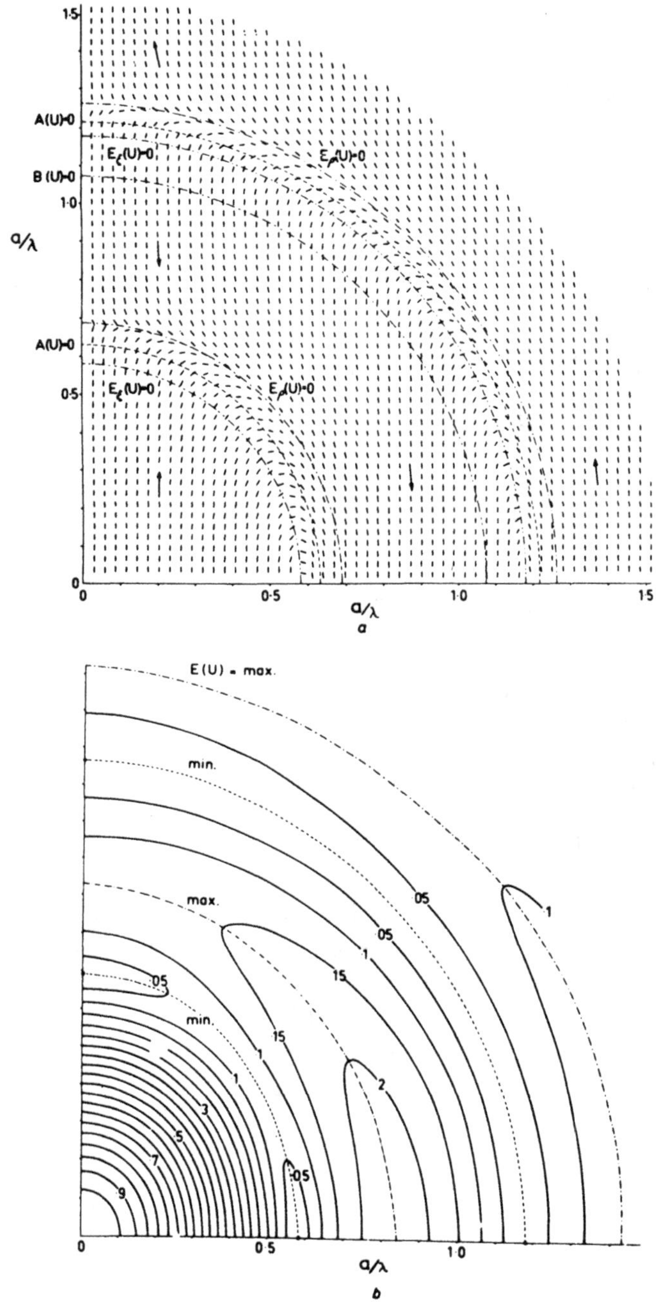

Fig. 2.5 *Field distribution in focal plane of paraboloid with $\theta_0 = 63°$ [176]*
a Polarisation of E field
b Contours of amplitude $E(u)$

a lens. For that case, observation showed the light intensity to comprise annular bands of light separated by dark rings with the intensity decreasing away from the lens axis. Even if we relax the constraint on θ_0 being small, the field in the central region remains nearly linearly polarised, and, as we shall see later, it matches well the field of the dominant HE_{11} hybrid mode. To effect a complete match to the focal field it would be necessary to synthesise all the cylindrical hybrid waves passing through the focal plane $z = 0$. This would require an infinitely overmoded waveguide which is practically unrealisable. For certain special applications where bandwidth is not a problem, a feed with several higher modes in addition to the dominant mode can be used to produce very high efficiency.

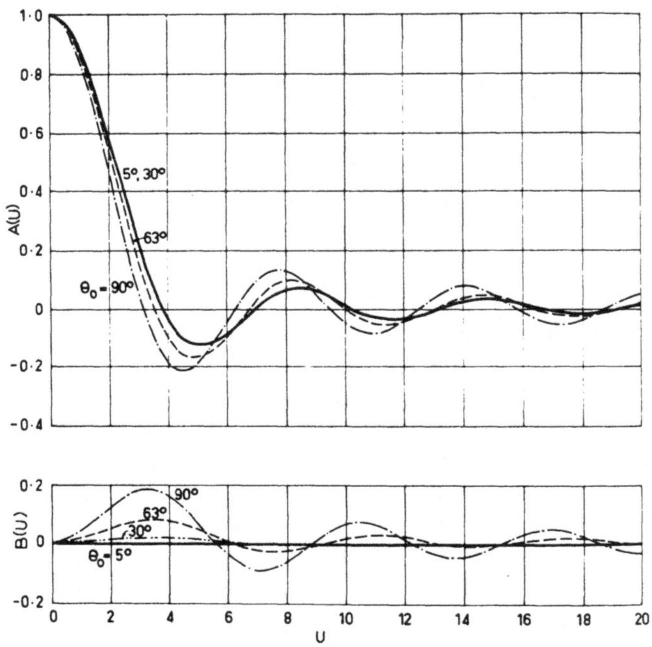

Fig. 2.6 *Functions A(U) and B(U) for paraboloid subtending semiangle θ_0 at focus [176]* *$A(U) = \alpha I_1$ and $B(U) = \alpha I_2$ are respectively proportional to the copolar and cross-polar focal field*

To better understand the relation between antenna efficiency and the focal field captured by the feed we will next examine the power flowing through the focal region. From eqns. 2.12 and 2.13 we can obtain the average rate of energy flow through an elemental area of the focal plane, S_z, as

$$S_z = \frac{(kfE_0)^2}{2Z_0} \left[|I_1|^2 - |I_2|^2\right] \tag{2.18}$$

where I_1 and I_2 are as in eqns. 2.14 and 2.15.

We next integrate S_z over an aperture in the focal plane of radius a, and introduce the parameter $U_a = ka \sin\theta_0$. Let the total power incident on the paraboloid be P_0,

$$P_0 = 2\pi\left(f\tan\frac{\theta_0}{2}\right)^2 \frac{E_0^2}{Z_0} \tag{2.19}$$

P/P_0 is then given by

$$\eta_0 = \frac{P}{P_0} = \frac{1}{2}\int_0^{U_a}\{A^2(U) - B^2(U)\}U\mathrm{d}U \tag{2.20}$$

where

$$A(U) = \frac{1}{2}\operatorname{cosec}^2\frac{\theta_0}{2} I_1 \tag{2.21}$$

$$B(U) = \frac{1}{2}\operatorname{cosec}^2\frac{\theta_0}{2} I_2 \tag{2.22}$$

and $U_a = ka \sin\theta_0$. $A(U)$ and $B(U)$ are plotted in Fig. 2.6.

For small θ_0, i.e. for a reflector with long focal length, $B(U) \cong 0$ and

$$\frac{P}{P_0} = 1 - J_0^2(U_a) - J_1^2(U_a) \tag{2.23}$$

which is a result derived originally by Lord Rayleigh.

Fig. 2.7 shows P/P_0 as a function of U_a. When θ_0 is small, the first maximum is obtained when $U_a \cong 3.8$, a value which corresponds to the first null of $A(U)$. For large θ_0, as U_a increases beyond this value, P/P_0 declines initially but this is not to say that the power coupled into an optimum hybrid-mode feed (shown as a chain curve in Fig. 2.7) will decline, this quantity in fact always increases monotonically. The explanation for this apparent paradox is that there are local regions off-axis where the direction of the Poynting vector is opposite to that on axis and these regions are larger for larger θ_0. However, as we now show, the fields that cause these regions of negative Poynting vector, do not couple to the field of a balanced hybrid mode and so the efficiency of an antenna fed by a waveguide supporting this dominant-mode field increases monotonically.

Let us rewrite eqns. 2.12 and 2.13 as

$$E_1 = E_A + E_B \tag{2.24}$$

$$H_1 = H_A + H_B \tag{2.25}$$

where

$$E_A = CA(U)\hat{i}_x \tag{2.26}$$

$$H_A = \frac{C}{Z_0} A(U)\,\hat{i}_y \tag{2.27}$$

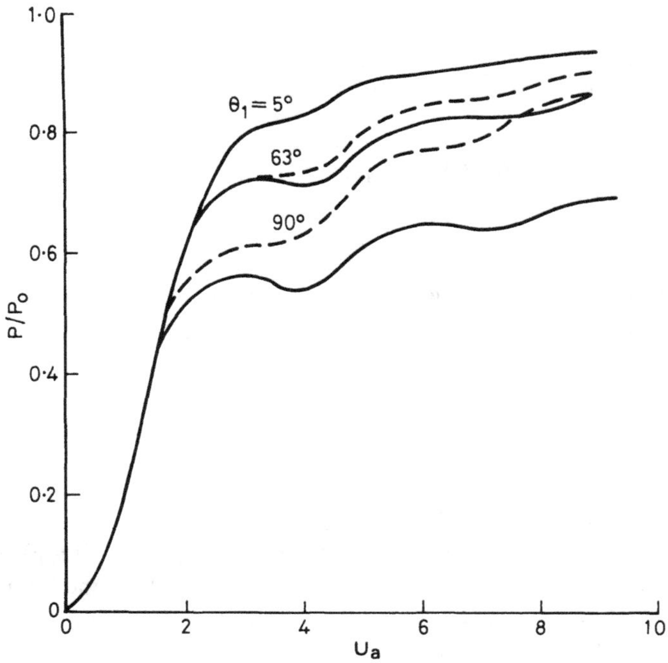

Fig. 2.7 *Normalised power flow in the focal plane of a paraboloidal reflector*
——— aperture efficiency
– – – – maximum aperture efficiency

$$E_B = CB(U)(\cos 2\phi_f \hat{i}_x + \sin 2\phi_f \hat{i}_y) \tag{2.28}$$

$$H_B = \frac{C}{Z_0} B(U)(\sin 2\phi_f \hat{i}_x - \cos 2\phi_f \hat{i}_y) \tag{2.29}$$

and C is a constant. Now consider a feed of aperture S carrying unit power, placed in the focal region with axis concident with that of the reflector and with fields E_2 and H_2 in its aperture. The power coupled into the feed by the focal fields E_1 and H_1 is given by

$$\eta = \frac{1}{2}\left|\int_S (\mathbf{E}_1 \times \mathbf{H}_2 + \mathbf{H}_1 \times \mathbf{E}_2)\cdot d\mathbf{S}\right|^2 \tag{2.30}$$

where it has been assumed that E_1 and H_1 arise from a plane wave with unit power incident on the reflector aperture. Now if

$$\mathbf{E}_2 = F(r_f,\phi)\hat{i}_y \tag{2.31}$$

and

$$\mathbf{H}_2 = -\frac{F}{Z_0}(r_f,\phi)\hat{i}_y \tag{2.32}$$

$$\eta = C\left|\int_0^{2\pi}\int_0^a A(U)F(r_f,\phi)r_f\mathrm{d}r_f\mathrm{d}\phi\right|^2 \tag{2.33}$$

Notice that in contrast with eqn. 2.20 a term in $B(U)$ does not arise and η increases monotonically with feed radius a. Maximum efficiency occurs when $F(r_f,\phi)\hat{i}_x$ matches $A(U)$ then

$$\eta_0 = \frac{1}{2}\int_0^{U_a} A(U)U\mathrm{d}U \tag{2.34}$$

Eqn. 2.34 is plotted in Fig. 2.7 where it can be seen that the corresponding value of η_0 is always greater than that of η_0 given by eqn. 2.20.

From eqn. 2.2 we see that when $X = Y$ (balanced-hybrid condition) the transverse field of the feed has a $J_0(Kr)$ dependance whereas $A(U)$ has the form $\{J_1(U)\}/U$ for small θ_0. However, if we arrange that the feed just encompasses the first null of $A(U)$, i.e. if we choose the radius r_1 such that

$$r_1 = \frac{3.83}{\sin\theta_0}\frac{\lambda}{\pi} \tag{2.35}$$

we will obtain maximum efficiency. The value for small θ_0 is

$$\begin{aligned}\eta_0 &= 1 - J_0^2(3.83) \\ &= 0.84\end{aligned} \tag{2.36}$$

Thus the maximum antenna efficiency of a front-fed paraboloid with a single-mode feed is 84%. Should we wish to improve on this value then we must improve the match between the focal field and the hybrid-mode field of the feed. We can do this either by using a dual-shaped reflector configuration or by a multimode feed. The former solution offers potentially a much wider bandwidth and is the solution chosen in, for example, all large earth-station antennas, whereas in radio-astronomical telescopes operating in the prime-focus mode of operation, the latter solution is sometimes chosen.

Although the symmetrical reflector of paraboloidal shape is most frequently used as an antenna, the offset paraboloid is gaining in importance. Offset configurations offer the advantage that the main reflector aperture is unblocked, the antenna efficiency is thus potentially higher, but, more important, there is minimal scattering of energy into the far-out sidelobes of the antenna pattern.

This makes the offset paraboloid of particular value as a space craft antenna and as an antenna for ultra-low sidelobe radars. We now ask whether the hybrid-mode feed remains the best choice for this type of antenna. The answer is 'yes' but with a reservation. Fig. 2.8 shows the focal field of an offset paraboloidal antenna with a long f/D. In contrast with the field of the symmetrical reflector shown in Fig. 2.5, we find that the offset reflector has two regions of high cross-polarisation falling within the central region of the focal field, consequently a conventional dominant mode hybrid feed will couple strongly to these fields and the cross-polar performance of the antenna is generally poor. All is not lost, however, for, as will be explained in more detail later, by adding an additional asymmetric hybrid mode to that of the HE_{11} mode, it is again possible to create a match to the focal field, restoring excellent cross-polar performance.

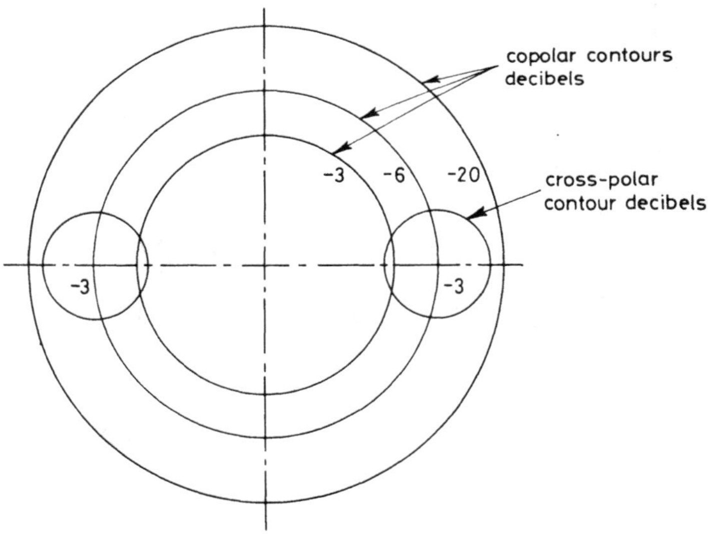

Fig. 2.8 *Approximate contour plot of typical focal-plane field distribution of an offset parabolic reflector uniformly illuminated from a distant linearly polarised source*

To complete this introduction, we show in Fig. 2.9 the field along the axis of a spherical reflector, as obtained by Thomas, Minnett and Vu [192]. The spherical reflector is useful in beam-scanning or multiple-beam applications and when the angle subtended at the paraxial focus is small (large f/D), there is appreciable energy confinement near the paraxial focus. Then a multimode hybrid feed placed at that position can couple efficiently to the focal field. Alternatively, by use of a Gregorian subreflector acting as an aberration corrector, see Fig. 2.10, Phillips and Clarricoats [187] have shown that efficient coupling to a hybrid-mode feed is possible over a wide range of subtended angles (f/D ratios). Although

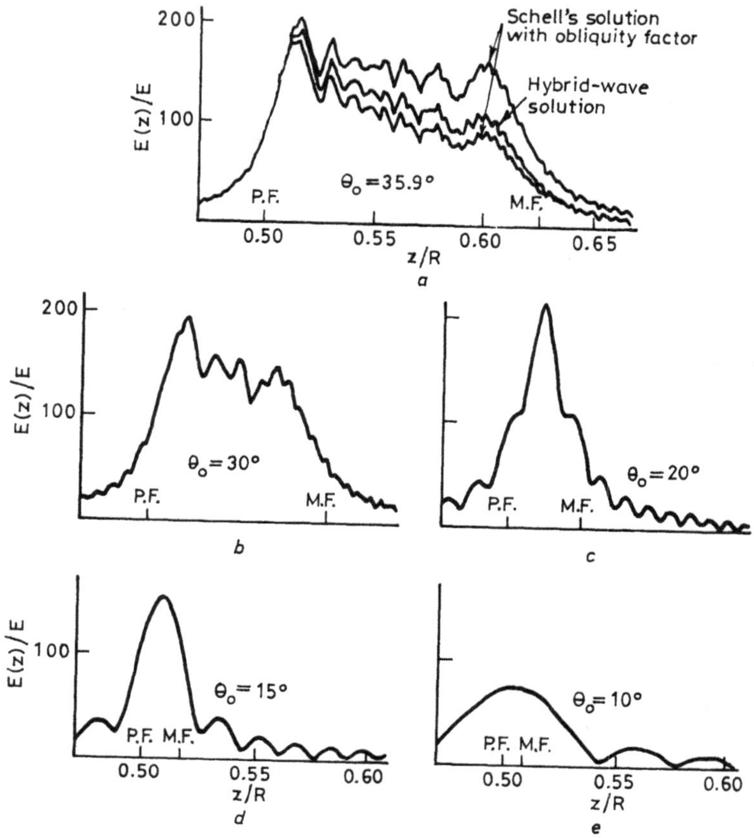

Fig. 2.9 *Normalised field along axis of a spherical reflector. R/λ = 400 for various values of* θ_0. *P.F. = paraxial focus. M.F. = marginal focus Thomas, Minnett and Vu [192]*

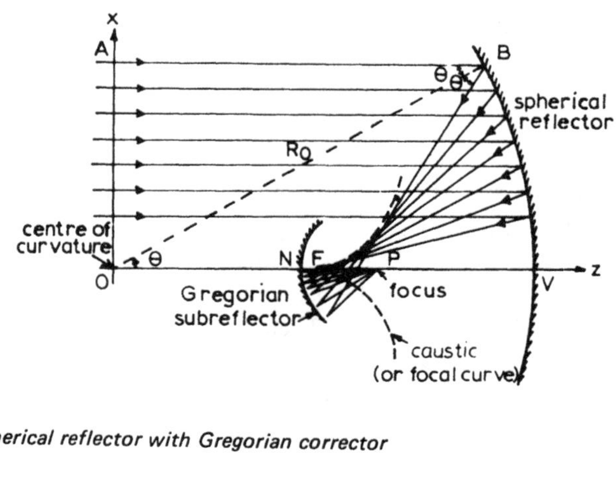

Fig. 2.10 *Spherical reflector with Gregorian corrector*

the spherical reflector offers advantages for beam scanning, the symmetric version of Fig. 2.10 suffers from serious blockage problems. An offset parabolic torus reflector, as investigated and produced by Hyde, overcomes that difficulty and it can be a very useful design for an earth station when communication with several satellites is required.

Chapter 3

Propagation and radiation characteristics of cylindrical corrugated waveguides

3.1 Introduction

The sucessful design of a feed with corrugated walls depends on the synergism of a number of related elements of theory. Fig. 3.1 identifies three main types of feed while Table 3.1 identifies the main elements of the theory.

In this chapter we are concerned with feeds in the class of Fig. 3.1 (a), i.e. the cylindrical corrugated waveguide with a radiating aperture. Feeds of this kind are used at the prime focus of a reflector for, as they can be made with aperture diameters as small as one wavelength or less, they provide for efficient illumination. The feed of Fig. 3.1 (a) also represents a good first approximation to the feed of Fig. 3.1 (b) provided that the horn flare angle is small and that a correction is made for the phase curvature of the aperture fields. Feeds of this kind are frequently used at the secondary focus of a Cassegrain antenna or in beam-waveguide fed antennas. Fig. 3.1 (c) shows a feed that combines features of both of the preceding types and draws on knowledge relevant to both. It is used when a narrow beamwidth radiation pattern is desired in a feed of short length.

Table 3.1 identifies the main elements of the theory required to determine the radiation pattern of a corrugated feed guide. First it is necessary to represent the boundary, we have two options, we can assume a uniform surface impedance for the wall or we can solve exactly the boundary-value problem. Then the field in the interior of the waveguide is represented by a space-harmonic series and the field within each corrugation by an infinite set of resonant modes. The first method is simple and quite accurate, however, when the waveguide radius is of order one wavelength, space harmonics must be included if an accurate crosspolar pattern is to be obtained. Additional computational time is involved in the space-harmonic solution. Next, in step two, an application of the boundary conditions leads to the modal propagation coefficients and transverse wavenumbers. In most cases we may assume the feed to be constructed from a perfect conductor and then the propagation coefficients are purely imaginary and the transverse wavenumbers purely real. Actually, the corrugated waveguide has interesting attenuation properties. When the walls have finite conductivity it transpires that the dominant HE_{11} mode exhibits lower attenuation than any mode of a uniform waveguide of comparable radius. We shall return to this feature later noting for the moment that an assumption of infinite conductivity has no effect on the radiation properties of the feed.

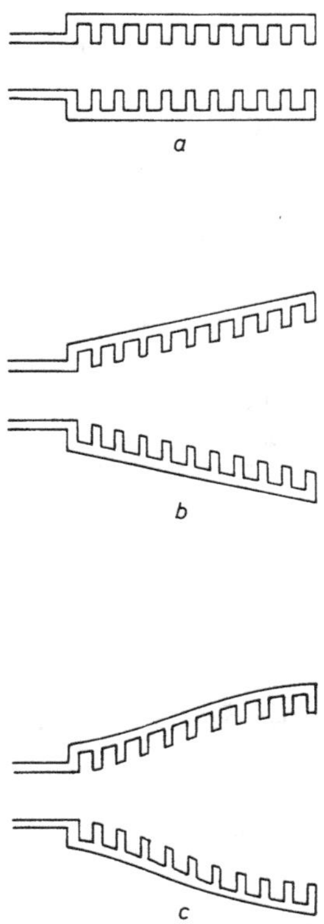

Fig. 3.1 *Types of cylindrical corrugated waveguide feeds*
a Open-ended waveguide
b Narrow flare angle horn
c Profiled horn

In step three, the transverse fields of the modes are expressed, and generally these are assumed to be the fields in the aperture. Actually at the aperture we are confronted with a complicated boundary value problem which has yet to be solved exactly. A complete solution would require the determination of the amplitudes and phases of all modes excited at the aperture together with a description of the currents which flow over the waveguide flange and along the outer wall of the waveguide. Fortunately the absence of a complete theory does not constitute a difficulty in design, for so far as the far-field radiation pattern is concerned, the finite flange can be treated as if it were infinite. This allows us in step four to determine the radiation pattern of a given mode most accurately by use of the Fourier Transform

Table 3.1 *Stages in theoretical determination of radiation pattern of corrugated feed*

1 Characterisation of feed boundary
   (a) Surface impedance
   (b) Inclusion of higher-order modes in slots and space harmonics in waveguide interior
2 Determination of modal propagation coefficients
3 Determination of fields in feed aperture
4 Determination of radiation fields by application of either:
   (a) Kirchhoff–Huygen method
   (b) Fourier–Transform method
   (c) Spherical wave expansion (conical horn)
   (d) Aperture integration with phase correction (conical horn)
5 Determination of transmission and reflection coefficient at horn mouth and (in case of conical horn) mode conversion along horn.

method. The Kirchhoff–Huygens method is also satisfactory for large apertures. If we desire we can correct for the finite size of the flange by using the Geometrical Theory of Diffraction, but the additional effort is seldom justified. Finally, in step five, we determine the amplitude and phase of the propagating mode or modes of the corrugated waveguide, together with the reflection coefficient of the incident mode in the input smooth-wall waveguide. This problem is solved by a straightforward modal expansion method.

We now develop the propagation equation and consider the mode characteristics combining, in effect, steps one to three above.

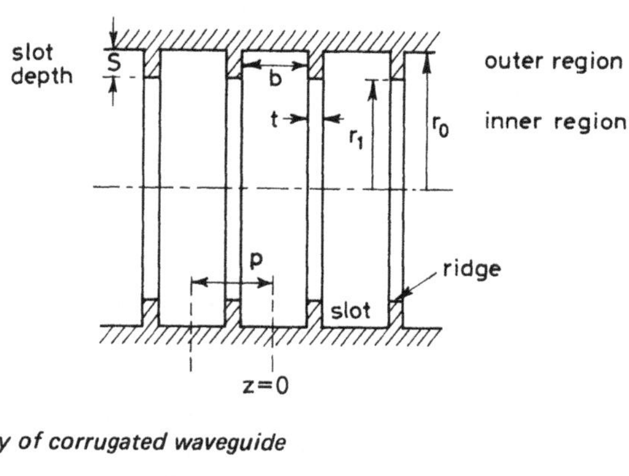

Fig. 3.2 *Geometry of corrugated waveguide*

3.2 Propagation in a cylindrical corrugated waveguide

Fig. 3.2 shows a cylindrical corrugated waveguide and defines the notation. In general, both longitudinal components of electric and magnetic field are required to satisfy the boundary conditions, the exceptions occurring at cut-off for all modes and for all values of the propagation coefficient, when the modes are azimuthally independent (TE_{0m} and TM_{0m} nodes). In the interior $r<r_1$ the longitudinal components of field E_z and H_z satisfy

$$\nabla_t^2 E_z + K_N^2 E_z = 0 \tag{3.1}$$

$$\nabla_t^2 H_z + K_N^2 H_z = 0 \tag{3.2}$$

where

$$K_N^2 = k^2 - \beta_N^2 \tag{3.3}$$

$$\beta_N = \beta_0 + \frac{2\pi N}{p} \tag{3.4}$$

$$k^2 = \omega^2 \epsilon_0 \mu_0 = (2\pi/\lambda_0)^2 \tag{3.5}$$

and a time dependence $e^{j\omega t}$ is assumed.

Solutions of eqns. 3.1 and 3.2 in cylindrical co-ordinates are

$$E_z = \sum_{N=-\infty}^{N=+\infty} A_{Nm} J_m(K_N r)\, e^{-j\beta_N z} \tag{3.6}$$

$$H_z = \sum_{N=-\infty}^{N=+\infty} B_{Nm} J_m(K_N r)\, e^{-j\beta_N z} \tag{3.7}$$

The form of the equations shows the fields to be expanded in space-harmonic travelling waves with propagation coefficients β_N related through the spatial period p, which is the separation distance of the corrugations.

On the perfectly conducting ridges between the annular slots E_z must vanish. Within the slots, with $r>r_1$,

$$E_z = \sum_{P=1}^{P=\infty} [C_P S_m^D \cos \eta_P z - jC_P' S^{D'} \sin \eta_P' z] \cos m\phi \tag{3.8}$$

$$H_z = \sum_{P=0}^{P=\infty} [-jD_P R_m^D \sin \eta_P z + D_P' R_m^{D'} \cos \eta_P' z] \sin m\phi \tag{3.9}$$

The form of the equations shows the fields within the slots to be standing waves and to satisfy the boundary conditions on the side walls of the slots,

$$\eta_p = 2\frac{P\pi}{b} \tag{3.10}$$

$$\eta_p' = (2P-1)\frac{\pi}{b} \tag{3.11}$$

In the above equations $J_m(x)$, $S_m(x, y)$, $R_m(x, y)$ are, respectively, Bessel and cylindrical functions of order m. The cylindrical functions are defined as follows:

$$R_m(x, y) = x \frac{J'_m(x)\, Y'_m(y) - J'_m(y)\, Y'_m(x)}{J_m(x)\, Y'_m(y) - J'_m(y)\, Y_m(x)} \tag{3.12}$$

$$= x \frac{R_m^N(x, y)}{R_m^D(x, y)} \tag{3.13}$$

$$S_m(x, y) = x \frac{J'_m(x) Y_m(y) - J_m(y)\, Y'_m(x)}{J_m(x)\, Y_m(y) - J_m(y)\, Y_m(x)} \tag{3.14}$$

$$= x \frac{S_m^N(x, y)}{S_m^D(x, y)} \tag{3.15}$$

To determine the propagation coefficient β_N and the form of the transverse fields required for the radiation pattern, boundary conditions at $r = r_1$ must be applied to the tangential components of the electric and magnetic fields, viz, E_z, E_ϕ, H_z and H_ϕ. In the Appendix, the propagation equation is formulated using space harmonics but here we shall proceed with a much simpler method, usually referred to as the surface-impedance approach. Notwithstanding this departure from rigour, from time to time, in discussion and in further developments, we shall draw upon the results of the space-harmonic formulation.

In the surface-impedance approach, we neglect the effect of the ridges and take only the lowest order TM_m standing wave in the slot which has components E_z, H_ϕ and H_r. This is a reasonable assumption provided the lowest order TE_m wave cannot be supported in the slot, i.e. the slot width $b < \lambda/2$. The approximation improves as the number of slots per wavelength increases and the thickness of the ridge decreases. The greatest effect is upon the ratio of H_z/E_z and this has the effect of altering the optimum design frequency as we shall see.

When space harmonics are neglected

$$E_z = a_{mn} J_m(Kr)\, e^{-j\beta z} \cos m\phi \tag{3.16}$$

$$H_z = b_{mn} J_m(Kr)\, e^{-j\beta z} \sin m\phi \tag{3.17}$$

For convenience, we drop the z dependence, and write eqns. 3.16 and 3.17 in the form

$$E_z = a_m J_m(x)\, e^{jm\phi} \tag{3.18}$$

$$H_z = -a_m j y_0 \bar{\Lambda} J_m(x)\, e^{jm\phi} \tag{3.19}$$

where $y_0 = (\epsilon_0/\mu_0)^{1/2}$ is the free-space wave admittance. $\bar{\Lambda}$ is the so-called normalised hybrid factor, a most important quantity in determining the crosspolar radiation characteristics of feeds. From Maxwell's equations we can express the ϕ components of the electric and magnetic fields.

For $r \leqslant r_1$

$$E_\phi = a_m \frac{x'}{x^2} J_m(x) \{m\bar{\beta} + \bar{\Lambda} F_m(x)\} e^{jm\phi} \tag{3.20}$$

$$H_\phi = -a_m \mathrm{j} y_0 \frac{x'}{x^2} J_m(x)\{\bar{\beta}\bar{\Lambda} + F_m(x)\}\mathrm{e}^{\mathrm{j}m\phi} \tag{3.21}$$

$$F_m(x) = x\frac{J'_m(x)}{J_m(x)} \tag{3.22}$$

$$x = Kr,\ x' = kr,\ \bar{\beta} = \frac{\beta}{k}$$

For $r_0 > r \geqslant r_1$, assuming only the TM_m standing wave in the slot,

$$E_z = \frac{a'_m}{Y_m(x'_0)} S_m^D(x', x'_0) \tag{3.23}$$

$$H_\phi = \frac{-a'_m \mathrm{j} y_0}{Y_m(x'_0)} S_m^N(x', x'_0) \tag{3.24}$$

$$x'_0 = kr_0$$

As there are no TE standing waves in the slot $E_\phi = 0$ at $r = r_1$, thus, from eqn. 3.20,

$$m\bar{\beta} = -\bar{\Lambda}F_m(x_1) \tag{3.25}$$

Eqn. 3.26 is just a special case of the general characteristic equation (29 in the Appendix). In eqn. 3.26, the fundamental space harmonic $N = 0$ is taken together with the first standing wave. Solutions of the exact characteristic equation differ The characteristic equation for β is obtained on applying continuity conditions to the surface admittance H_ϕ/E_z at $r = r_1$.

$$F_m(x_1) - \frac{(m\bar{\beta})^2}{F_m(x_1)} = \left\{\frac{x_1}{x'_1}\right\}^2 S_m(x'_1, x'_0) \tag{3.26}$$

from those of eqn. 3.26 in ways we shall discuss but many general features of the exact solutions of Fig. 3.3 are the same. The sketches of Fig. 3.4, which correspond to different values of r_1/r_0, and Figs. 3.5–3.7, help us to identify these features as follows.

The cut-off condition $\bar{\beta} = 0$

In eqn. 3.26 the condition $\bar{\beta} = 0$ yields either

$$F_m(x_1) = 0 \tag{3.27}$$

or

$$F_m(x_1) = S_m(x'_1, x'_0) \tag{3.28}$$

Eqn. 3.27 is satisfied when

$$J'_m(x_1) = 0 \tag{3.29}$$

then H_z is finite and $E_z = 0$. The cut-off is therefore an H-type and in the limit as $\bar{\beta}$ tends to be zero, the mode exhibits pure TE properties with $\bar{\Lambda}$ tending to infinity.

Eqn. 3.28 is satisfied when

$$J_m(x_0) = 0 \tag{3.30}$$

then E_z is finite and $H_z = 0$. The cut-off is therefore an E-type in the limit as $\bar{\beta}$ tends to zero, the mode exhibits pure TM properties with $\bar{\Lambda}$ tending to zero. Notice that for H-mode cut-off the boundary condition corresponds to that of a TE mode in a uniform waveguide of radius r_1 while for E-mode cut-off the boundary condition corresponds to that of a TM mode in a uniform waveguide of radius r_0. The difference arises because H_z is always zero for $r > r_1$ subject to our assumption about the absence of TE waves in the slot, while by contrast E_z can penetrate the slot to vanish on the perfectly conducting base of the slot at $r = r_0$.

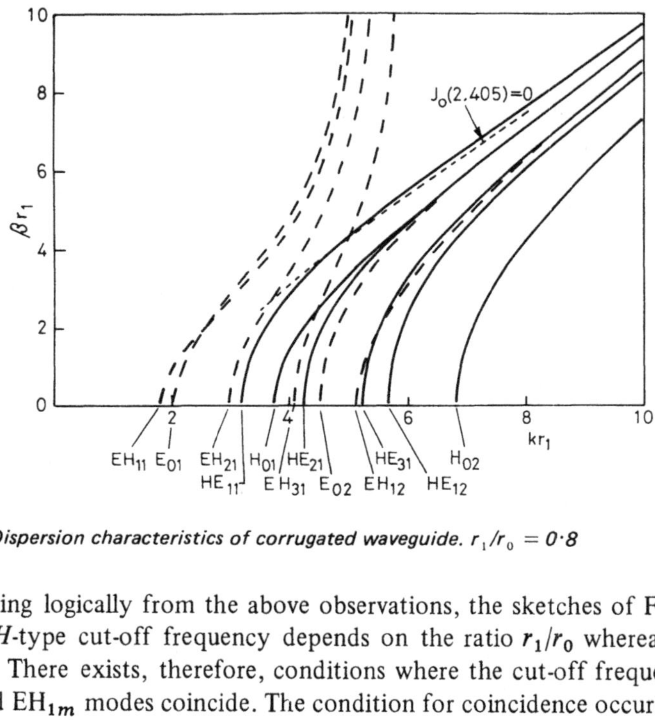

Fig. 3.3 *Dispersion characteristics of corrugated waveguide.* $r_1/r_0 = 0{\cdot}8$

Following logically from the above observations, the sketches of Fig. 3.4 show that the H-type cut-off frequency depends on the ratio r_1/r_0 whereas the E-type does not. There exists, therefore, conditions where the cut-off frequencies of the HE_{1m} and EH_{1m} modes coincide. The condition for coincidence occurs when

$$F_1(x_1) = S_1(x_1', x_0') = 0 \tag{3.31}$$

For higher order $n = 1$ modes, this occurs for higher values of the ratio r_1/r_0 and then, for very large values of kr_1, from the properties of Bessel functions, $k(r_0 - r_1) = \pi$ in the limit. Thus the degeneracy conditions occurs for high-order modes when the corrugation depth is approximately $\lambda/2$.

When designing corrugated horns, especially those with narrow flare angle, we are interested in the way the special conditions vary with the normalised corrugation depth $(r_0 - r_1)/\lambda$ as kr_1 varies. This information is plotted in Fig. 3.6.

The balanced-hybrid conditions $S_m(x_1', x_0') = 0$

When the surface admittance $Y = 0$, $S_m(x_1', x_0') = 0$ and we have from eqn. 3.26

$$F_m^2(x_1) = (m\bar{\beta})^2 \tag{3.32}$$

which together with eqn. 3.25 leads to

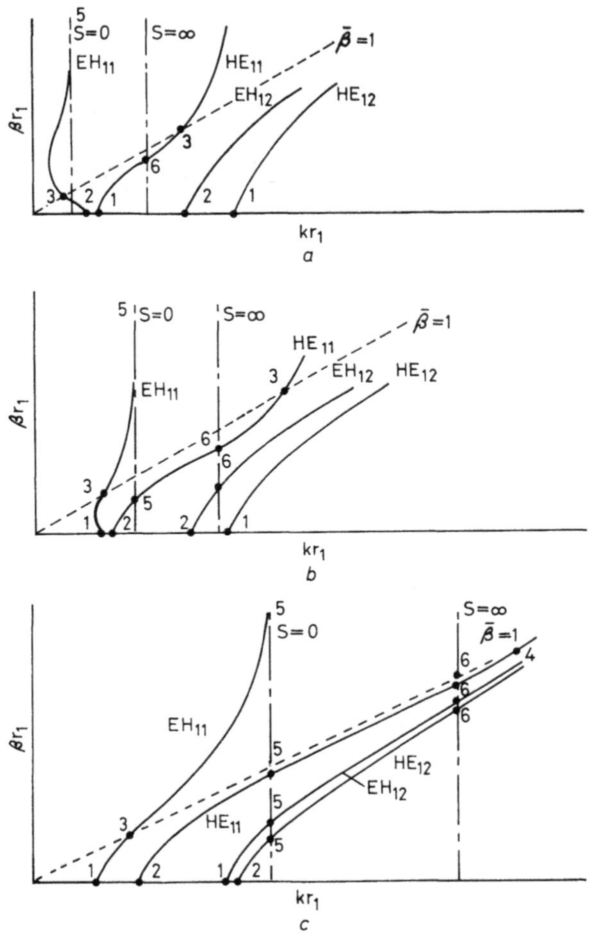

Fig. 3.4 *Dispersion diagrams for corrugated waveguide*
a $r_1/r_0 = 0{\cdot}4$
b $r_1/r_0 = 0{\cdot}63$ (typical of the throat of a horn)
c $r_1/r_0 = 0{\cdot}8$ (typical of the aperture of a horn)
Points 1 and 2 represent cut-off for modes with $E_z = 0$ and $H_z = 0$, respectively, at $\beta = 0$
Points 3 and 4 represent condition $\bar{\beta} = \beta/k = 1$, corresponding to $S_1(k, r_1, r_0) = -\frac{1}{2}(kr_1)^2$
Point 5 represents balanced hybrid or 'open-circuit' boundary condition corresponding to $S_1(k, r_1, r_0) = 0$
Point 6 represents short-circuit boundary condition corresponding to $S_1(k, r_1, r_0) = \infty$

$$(\bar{\Lambda})^2 = 1 \tag{3.33}$$

We refer to this as the balanced-hybrid condition since eqns. 3.18 and 3.19 give when $\bar{\Lambda} = 1$

$$\left|\frac{E_z}{H_z}\right| = \left(\frac{\mu_0}{\epsilon_0}\right)^{1/2} \qquad r < r_1 \tag{3.34}$$

Eqn 3.33 has two roots,

$$\bar{\Lambda} = +1 \tag{3.35}$$

corresponding to HE modes and

$$\bar{\Lambda} = -1 \tag{3.36}$$

corresponding to EH modes.

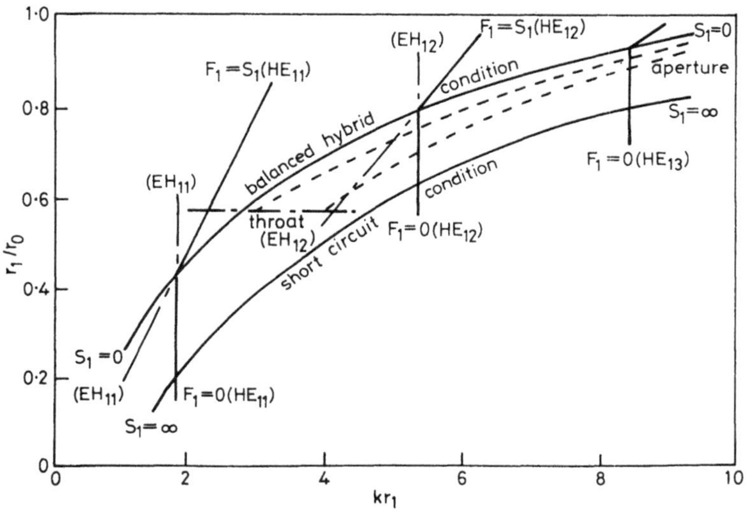

Fig. 3.5 *Parametric dependence of special points as function of r_1/r_0 chain curves denote two examples of loci for HE_{11} mode in a flared horn.*

The condition is marked as 5 in Figs. 3.4 and 3.7. Fig. 3.7 shows $\bar{\Lambda}$ as a function of kr_1 for values of r_1/r_0 corresponding to Fig. 3.4. We have seen previously that $\bar{\Lambda}$ equals 0 or ∞ at cut-off depending on the type of cut-off. In Fig. 3.7(c), r_1/r_0 is such that all the HE_{1m} modes shown have E-type cut-off for which $\bar{\Lambda} = 0$. Then as kr_1 increases, $\bar{\Lambda}$ increases to reach the balanced condition $\bar{\Lambda} = +1$ before tending to infinity at the 'short-circuit' condition where $\bar{\Lambda} = S_1 = \infty$ and $r_0 - r_1 \simeq \lambda/2$. We note that for asymptotically large values of kr_1, not represented in Figs. 3.4–3.7, $\bar{\beta}$ tends to unity at the condition

$$J_q(x_1) = 0 \tag{3.37}$$

where $q = 0$ for HE_{1m} modes and $q = 2$ for EH_{1m} modes. The lowest order solution for the HE_{11} mode occurs when

$$x_1 = kr_1 = 2{\cdot}405 \tag{3.38}$$

Dragone [56] has pointed out that, within the surface-impedance model, for larger

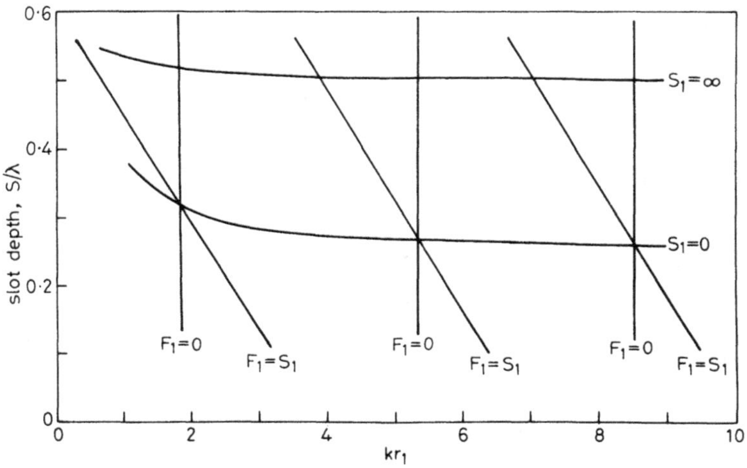

Fig. 3.6 *Parametric dependence of special points as function of normalised slot depth*

Table 3.2

For HE_{1m} modes

$$x_1 = x_{0m}\left\{1 - \frac{Y}{2kr_1} - \tfrac{1}{2}\left[1 - \frac{Y^2}{4}(1 + x_{0m}^2)\right]\frac{1}{(kr_1)^2} + \frac{Y}{4}\left[1 - \frac{Y^2}{12}(7x_{0m}^2 + 1)\right]\frac{1}{(kr_1)^3}\ \cdots\right\} \tag{3.39}$$

where x_{nm} is the mth root of $J_n(x) = 0$

$$\bar{\Lambda} = 1 - x_{0m}^2\left\{\frac{Y}{2kr_1} - \frac{Y^2}{8}(4 + x_{0m}^2)\frac{1}{(kr_1)^2} - \frac{Y}{2}\left[1 - \frac{x_{0m}^2}{2} - \frac{Y^2}{4}(3x_{0m}^2 + 2)\right]\frac{1}{(kr_1)^3}\ \cdots\right\} \tag{3.40}$$

For EH_{1m} modes

$$x'_m = x_{2m}\left\{1 - \frac{Y}{2kr_1}\ \cdots\right\} \tag{3.41}$$

$$\bar{\Lambda} = -1 - x_{2m}^2\,\frac{Y}{2kr_1}\ \cdots \tag{3.42}$$

and larger values of kr_1, the actual value of the surface impedance has less and less effect on the propagation behaviour of a fast hybrid mode. Eqn. 3.38 generally represents the lowest order asymptotic solution of eqn. 3.26 for HE_{1m} modes.

Table 3.2, containing eqns. 3.39–3.42 due to Dragone [56], gives general asymptotic series for x_1 and $\bar{\Lambda}$. Y is the normalised surface admittance of eqn. 2.4.

The short-circuit condition $S_m(x_1', x_0') = \infty$

The next special condition with increasing kr_1 occurs for fixed r_1/r_0 when $S = \infty$ and the corrugation depth is approximately one half-wavelength. The surface impedance is then zero. For this case,

$$Y = S_m(x_1', x_0') = \infty \tag{3.43}$$

corresponding to the label 6 in Figs. 3.4 and 3.7. Under this condition the waveguide behaves as if it had a continuous perfectly conducting boundary at $r = r_1$. Pure mode solutions are obtained with

$$E_z = 0 \text{ for TM modes at } r = r_1$$

or

$$\frac{dH_z}{dr} = 0 \text{ for TE modes at } r = r_1$$

The condition is important in the design of the throat of a corrugated horn.

The fast-wave to slow-wave transition point $\bar{\beta} = 1$

Next, for a fixed value of r_1/r_0, as $\bar{\beta}$ is increased we encounter the fast-wave to slow-wave transition at $\beta = 1$. When $\beta = 1$, $K = 0$ and $F_m(x) = \pm m$. $\bar{\Lambda} = -1$ the point marked ③ while $\bar{\Lambda} = +1$ the point marked ④. From an expansion of eqn. 3.26 about $x_1 = 0$, we have for $m = 1$

$$S_1(x_1', x_0') = -\tfrac{1}{2}(kr_1)^2 \tag{3.44}$$

The high-frequency cut-off $\bar{\beta} = \infty$

Subject to the surface-impedance approximation, as frequency increases within the slow-wave domain, the condition is eventually reached where $\bar{\beta}$ tends to infinity. This is known as high-frequency cut-off. It corresponds to the condition $\beta p = \pi$ for a periodic structure, with p, the period, tending to zero. Consider $m = 1$ in eqn. 3.26. The asymptotic form of $F_1(ja)$ with (a) real is

$$F_1(ja) = a \tag{3.45}$$

then in the limit the condition requires

$$S_1(x_1', x_0') = 0 \tag{3.46}$$

Thus as Fig. 3.4(c) shows, as the HE_{11} reaches a balanced hybrid condition with

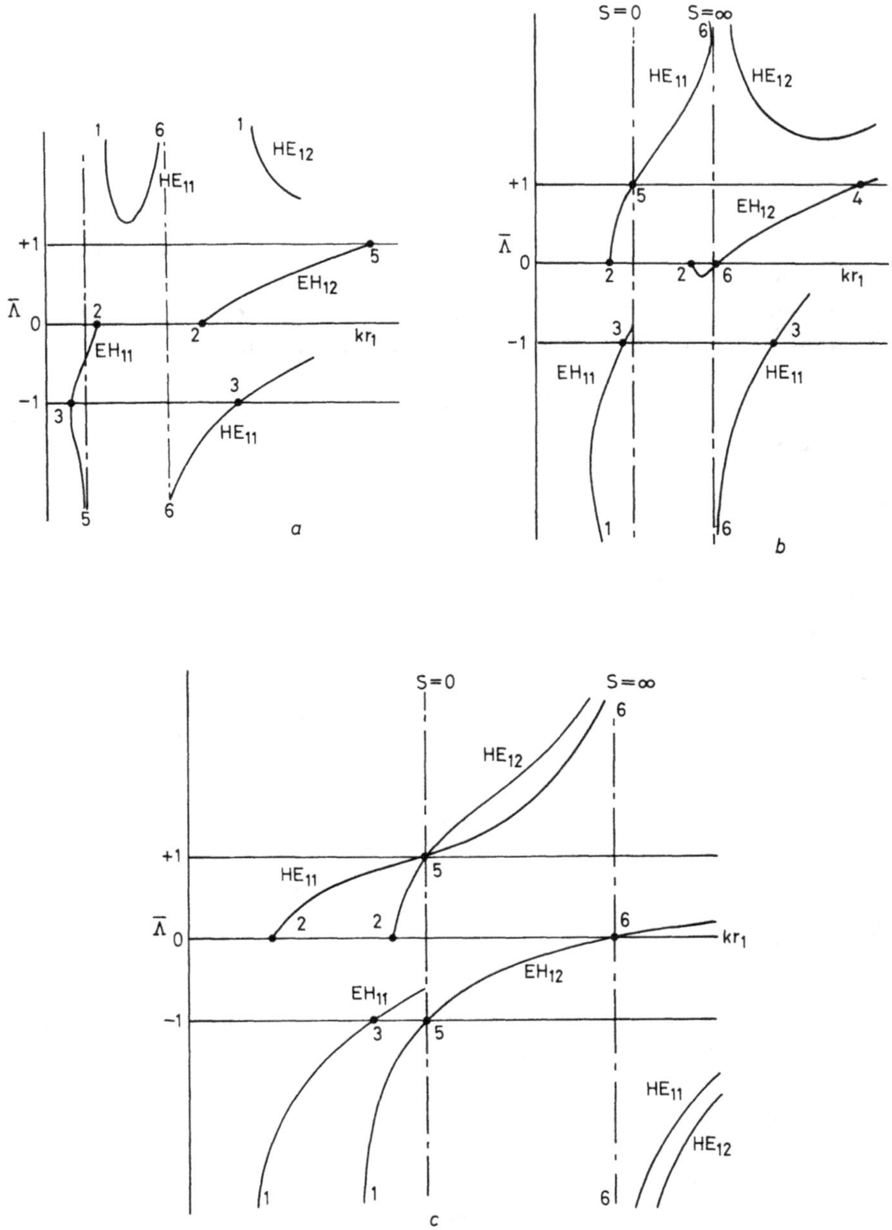

Fig. 3.7 *Hybrid factor diagrams corresponding to dispersion diagrams of Fig. 3.4*
a $r_1/r_0 = 0{\cdot}4$
b $r_1/r_0 = 0{\cdot}63$
c $r_1/r_0 = 0{\cdot}8$

$\bar{\Lambda} = +1$ the slow EH_{11} mode escapes to high-frequency cut-off. We also note that at high-frequency cut-off the value of $\bar{\Lambda}$ is given by

$$\bar{\Lambda} = -\frac{1}{kr_1} \tag{3.47}$$

Summary

Let us summarise the behaviour as kr_1 increases from cut-off with r_1/r_0 constant at a value greater than 0·75, using Figs. 3.4(c) and 3.7(c) as a guide. For HE_{1m} modes, cut-off has $\bar{\beta} = 0$ and $\bar{\Lambda} = 0$. The balanced-hybrid condition has $\bar{\Lambda} = +1$, and for large kr_1 the corrugation depth approaches $\lambda/4$. Beyond this as we approach the short-circuit condition (corrugation depth $\lambda/2$), the mode tends to a pure TE mode and $\bar{\Lambda} = +\infty$.

The mode continues with $\bar{\Lambda} = -\infty$ and as we approach the fast-wave to slow-wave transition, point $\bar{\beta} = 1$, we have now $\bar{\Lambda} = -1$. Finally, as frequency increases further the mode terminates at high-frequency cut-off with $\bar{\beta} = \infty$ and the corrugation depth is $3\lambda/4$, then $\bar{\Lambda} = -1/kr_1$ which tends to zero as kr_1 tends to infinity. We see that as $\bar{\beta}$ progresses from 0 to ∞, $\bar{\Lambda}$ progresses from 0 to $+\infty$ and from $-\infty$ back to near zero again.

For EH_{1m} modes, we begin at $\bar{\beta} = 0$ with $\bar{\Lambda} = -\infty$ and pure TE conditions. For the lowest order EH_{1m} mode, the pass-band is narrow and the wave terminates with $\bar{\beta} = \infty$ at the same point that the HE_{11} mode reaches the balanced-hybrid condition. We have termed this first EH_{1m} mode the EH_{11} mode, but some authors, for example, Dragone [56], simply refer to this mode as 'the surface wave mode'. Such nomenclature is ambiguous for obvious reasons. The next EH_{1m} mode, our EH_{12}, remains a fast wave through the first balanced-hybrid condition where the corrugation depth is approximately $\lambda/4$, it has $\bar{\Lambda} = 0$ and behaves as a pure TM mode when the corrugation depth is $\lambda/2$, it becomes a slow wave when $\bar{\Lambda} = +1$ and mode terminates when the corrugation depth is $3\lambda/4$. We now see that in general, within the surface-impedance model, a mode is sustained with increasing frequency over the range in which the corrugation depth increases by slightly more than $\lambda/2$. The behaviour of β and $\bar{\Lambda}$ for r_1/r_0 near and below the degenerate cut-off is also shown in Figs. 3.4 and 3.7. A detailed account will not be presented but it should be recognised that such values of r_1/r_0 can prevail in the throat of a horn.

Effect of space harmonics

Fig. 3.3 shows βr_1 as a function of kr_1 with inclusion of the $N = \pm 1$ space harmonics in addition to the fundamental; higher harmonics beyond the $N = \pm 1$ pair have negligible effect on βr. Inclusion of harmonics has an influence on all of the special conditions discussed above to an extent which depends on the mode and the waveguide parameters.

Chong [23] has made a detailed comparison for the TE_{01}, TM_{01}, EH_{11} and HE_{11} modes but because the latter mode is the most important in feed applications we concentrate on his results for the HE_{11} mode. Table 3.3 shows a comparison of

Table 3.3 *Comparison of low frequency cut-off values (kr_1) of the HE_{11} mode*

r_1/r_0	0·8	0·7	0·6	0·5	0·4
SI model	3·0654	2·6822	2·2990	1·9159	1·5327
SH model					
at $\beta_0 r_1 = 0{\cdot}1$	3·1108	2·7215	2·3202	1·9221	1·7742
% error	1·46	1·44	0·91	0·33	13·6
$\bar{\Lambda}$ at $\beta_0 r_1 = 0{\cdot}1$	0·00233	0·0124	0·0372	0·239	31·7

the surface impedance (SI) and space-harmonic (SH) models, with r_1/r_0 as parameter, of the normalised frequency at low-frequency cut-off ($\beta = 0$). Actually, because of a singularity at $\beta = 0$, the comparison is made for a value of $\beta r_1 = 0{\cdot}1$ rather than $\beta r_1 = 0$; this small difference has a quite negligible effect on the normalised frequency because $\partial\beta/\partial\omega$ is very large near cut-off. The percentage error in using the SI model is seen to be less than 2% in the range $0{\cdot}5 \leqslant r_1/r_0 \leqslant 0{\cdot}8$. For the other modes referred to above, the maximum error in the same range is about 4%, thus we can use the SI cut-off mode chart of Fig. 3.8 with confidence. Table

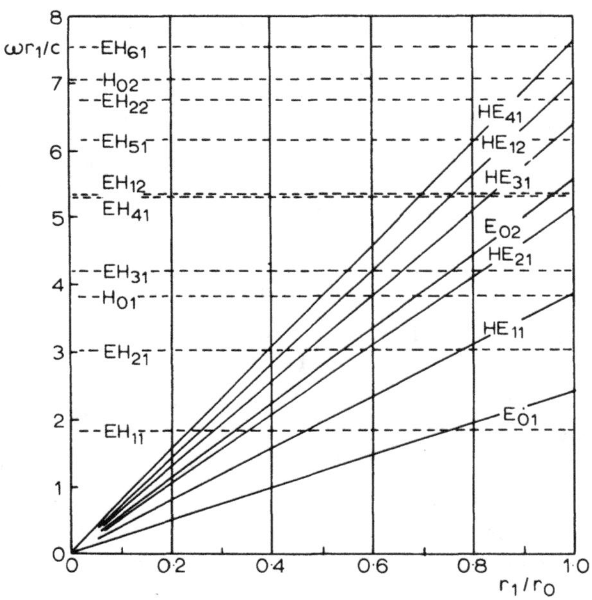

Fig. 3.8 *Low-frequency cut-off mode chart*

Because of the degeneracy of cut-off discussed in the text the nomenclature refers to cut-off for r_1/r_0 values greater than the degeneracy point. Below this the nomenclature is reversed.

3.4 makes a similar comparison for the high-frequency cut-off. Here the agreement between the models is weaker because, at the high-frequency cut-off, there is equal power in the fundamental and $N = -1$ harmonics so neglect of space harmonics

Table 3.4 *Comparison of high frequency cut-off values (kr_1) of the HE_{11} mode*

r_1/r_0	0·7	0·6	0·5	0·4
SI model	11·23	7·32	4·85	3·29
SH model	–	6·5	4·5	3·17
% error	–	12·6	7·8	3·8

in the SI model is clearly less valid. Nevertheless, the mode chart of Fig. 3.9 is a useful guide to design.

In subsequent discussion of the radiation characteristics of corrugated waveguides and horns we shall frequently make use of the SI model in the vicinity of $\bar{\Lambda} = 1$.

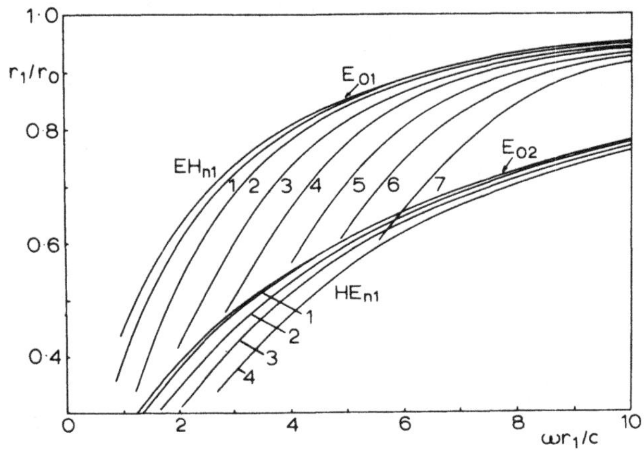

Fig. 3.9 *High-frequency cut-off mode chart*

Fig. 3.10 compares for the HE_{11} mode, the normalised wave number Kr_1 as a function of the normalised frequency kr_1 using the two models. Corresponding values of $\bar{\Lambda}$ are shown in the vicinity of $\bar{\Lambda} = 1$. The value of kr_1 at $\bar{\Lambda} = 1$ differs by 10% between the two models for this value of r_1/r_0. Fig. 3.10 also allows us to check the assumption $J_0(Kr_1) = 0$ which is sometimes used as an approximation to characterise fields in and radiation from a corrugated waveguide. In Fig. 3.3 we include points on the propagation curve obtained by making this simple assumption where it can be seen that over a useful range of values away from cut-off, the approximation is quite good. Fig. 3.10 does show, however, that the approximation is rather worse than the SI model might lead one to suppose.

To complete this discussion of the propagation characteristics of corrugated waveguides Fig. 3.11 shows for the case $r_1/r_0 = 0{\cdot}6$, $\bar{\beta}$ and $\bar{\Lambda}$ for the four lowest HE_{1m} modes. The influence of the finite period is manifest among the higher order modes where the high-frequency cut-off occurs for a value kr_1 close to that at which $\bar{\beta}$ is unity. The condition $\bar{\Lambda} = 1$ occurs close to the frequency at which the corrugation depth is a multiple of $\lambda/4$.

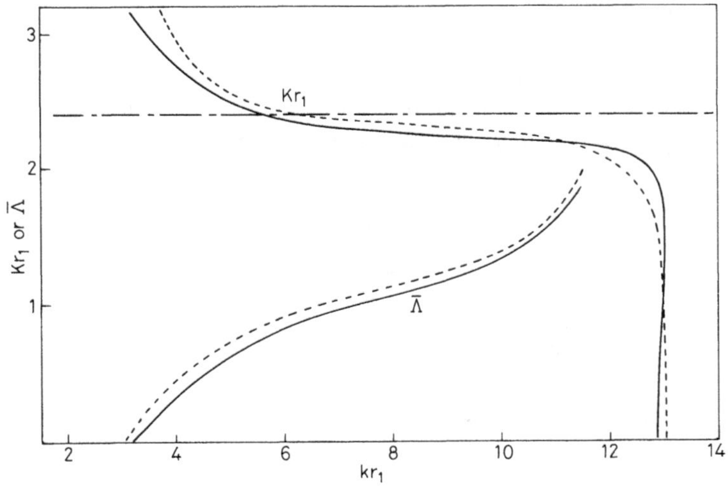

Fig. 3.10 *Comparison of wavenumbers and hybrid factor in space harmonic and surface impedance calculations: $r_1/r_0 = 0{\cdot}8$, $s/r_1 = 0{\cdot}2$, $t/b = 0{\cdot}3$*

——— space harmonic calculation

– – – surface impedance calculation

—·— $Kr_1 = 2{\cdot}405$

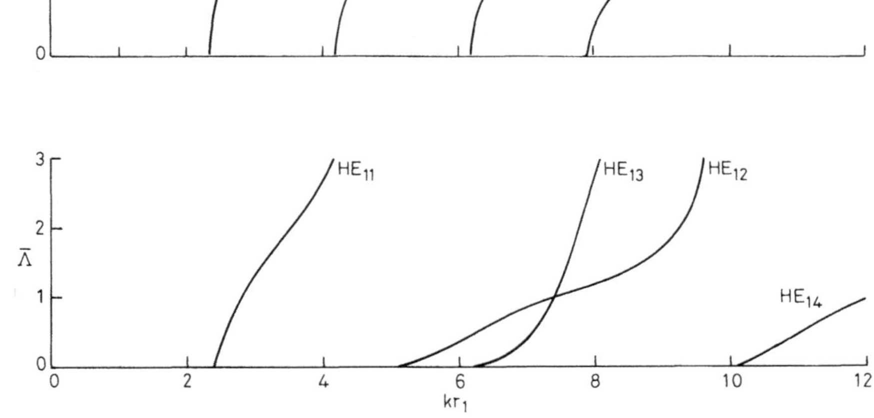

Fig. 3.11 *Dispersion characteristics and hybrid factor for HE_{1m} modes: $r_1/r_0 = 0{\cdot}6$, $s/r_1 = 0{\cdot}2$, $t/b = 0{\cdot}3$, $K = L = 1$*

3.3 Fields in a cylindrical corrugated waveguide

Our main objective, the radiation characteristics of hybrid-mode feeds, requires as an intermediate goal the fields in the feed aperture. For the cylindrical waveguide, subject to the use of the SI model, the electric and magnetic fields are obtained using eqns. 3.18–3.21. Table 3.5 presents these equations in polar and cartesian form, numbered 3.48–3.62.

Table 3.5

$$E_z = a_m J_m(x)\cos m\phi \tag{3.48}$$

$$H_z = a_m y_0 \bar{\Lambda} J_m(x)\sin m\phi \tag{3.49}$$

$$E_r = -\mathrm{j}a_m \frac{k}{K^2} J_m(x)[\bar{\beta}F_m(x) + m\bar{\Lambda}]\cos m\phi \tag{3.50}$$

$$H_r = -\mathrm{j}a_m y_0 \frac{k}{K^2} J_m(x)[\bar{\beta}\bar{\Lambda}F_m(x) + m]\sin m\phi \tag{3.51}$$

$$E_\phi = \mathrm{j}a_m \frac{k}{K^2} J_m(x)[m\bar{\beta} + \bar{\Lambda}F_m(x)]\sin m\phi \tag{3.52}$$

$$H_\phi = -\mathrm{j}a_m y_0 \frac{k}{K^2} J_m(x)[m\bar{\beta}\bar{\Lambda} + F_m(x)]\cos m\phi \tag{3.53}$$

$$F_m(x) = xJ'_m(x)/J_m(x),\ x = Kr. \tag{3.54}$$

$$E_r = -\mathrm{j}\frac{a_1}{2}\frac{k}{K}[(\bar{\beta} + \bar{\Lambda})J_0(x) + (\bar{\Lambda} - \bar{\beta})J_2(x)]\cos\phi \tag{3.55}$$

$$E_\phi = \mathrm{j}\frac{a_1}{2}\frac{k}{K}[(\bar{\beta} + \bar{\Lambda})J_0(x) - (\bar{\Lambda} - \bar{\beta})J_2(x)]\sin\phi \tag{3.56}$$

$$H_r = -\mathrm{j}\frac{a_1}{2}y_0\frac{k}{K}[(\bar{\beta}\bar{\Lambda} + 1)J_0(x) + (1 - \bar{\beta}\bar{\Lambda})J_2(x)]\sin\phi \tag{3.57}$$

$$H_\phi = -\mathrm{j}\frac{a_1}{2}y_0\frac{k}{K}[(\bar{\beta}\bar{\Lambda} + 1)J_0(x) - (1 - \bar{\beta}\bar{\Lambda})J_2(x)]\cos\phi \tag{3.58}$$

$$E_x = -\mathrm{j}\frac{a_1}{2}\frac{k}{K}[(\bar{\beta} + \bar{\Lambda})J_0(x) + (\bar{\Lambda} - \bar{\beta})J_2(x)\cos 2\phi] \tag{3.59}$$

$$E_y = -\mathrm{j}\frac{a_1}{2}\frac{k}{K}(\bar{\Lambda} - \bar{\beta})J_2(x)\sin 2\phi \tag{3.60}$$

$$H_x = -\mathrm{j}y_0\frac{a_1}{2}\frac{k}{K}(1 - \bar{\Lambda}\bar{\beta})J_2(x)\sin 2\phi \tag{3.61}$$

$$H_y = -\mathrm{j}y_0\frac{a_1}{2}\frac{k}{K}[(1 + \bar{\beta}\bar{\Lambda})J_0(x) - (1 - \bar{\beta}\bar{\Lambda})J_2(x)\cos 2\phi] \tag{3.62}$$

The case $m = 1$ is of particular interest and in equations 3.59–3.62 we present the corresponding Cartesian components of the tranverse electric and magnetic fields. Two features are evident; first, the condition $\bar{\Lambda} = \bar{\beta}$ renders the transverse electric field E_x independent of ϕ and the component E_y identically zero. This

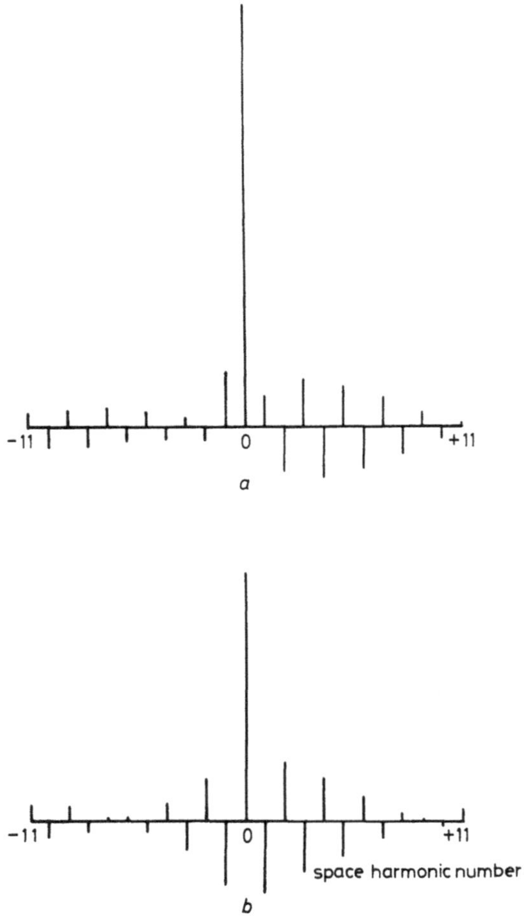

Fig. 3.12 *Amplitude of space harmonic coefficients for HE_{11} mode: $K = L = 11$, $r_1/r_0 = 0{\cdot}8$, $r_1 = 40$ mm, $s = 10$ mm, $t/b = 0{\cdot}1$, $kr_1 = 8{\cdot}4$*
a Electric field coefficient A
b Magnetic field coefficient B

condition is important because when the Fourier Transform method is used to obtain the radiation field from the aperture electric field, the same condition yields zero radiated crosspolarisation when the radiating aperture is large in terms of wavelength. Secondly, we note that the condition which renders the transverse magnetic field to be linearly polarised is not $\bar{\Lambda} = \bar{\beta}$, but $\bar{\Lambda}\bar{\beta} = 1$. The two conditions occur at nearly the same frequency when $\bar{\beta} \rightarrow 1$ which it does for a large aperture waveguide in the SI modes.

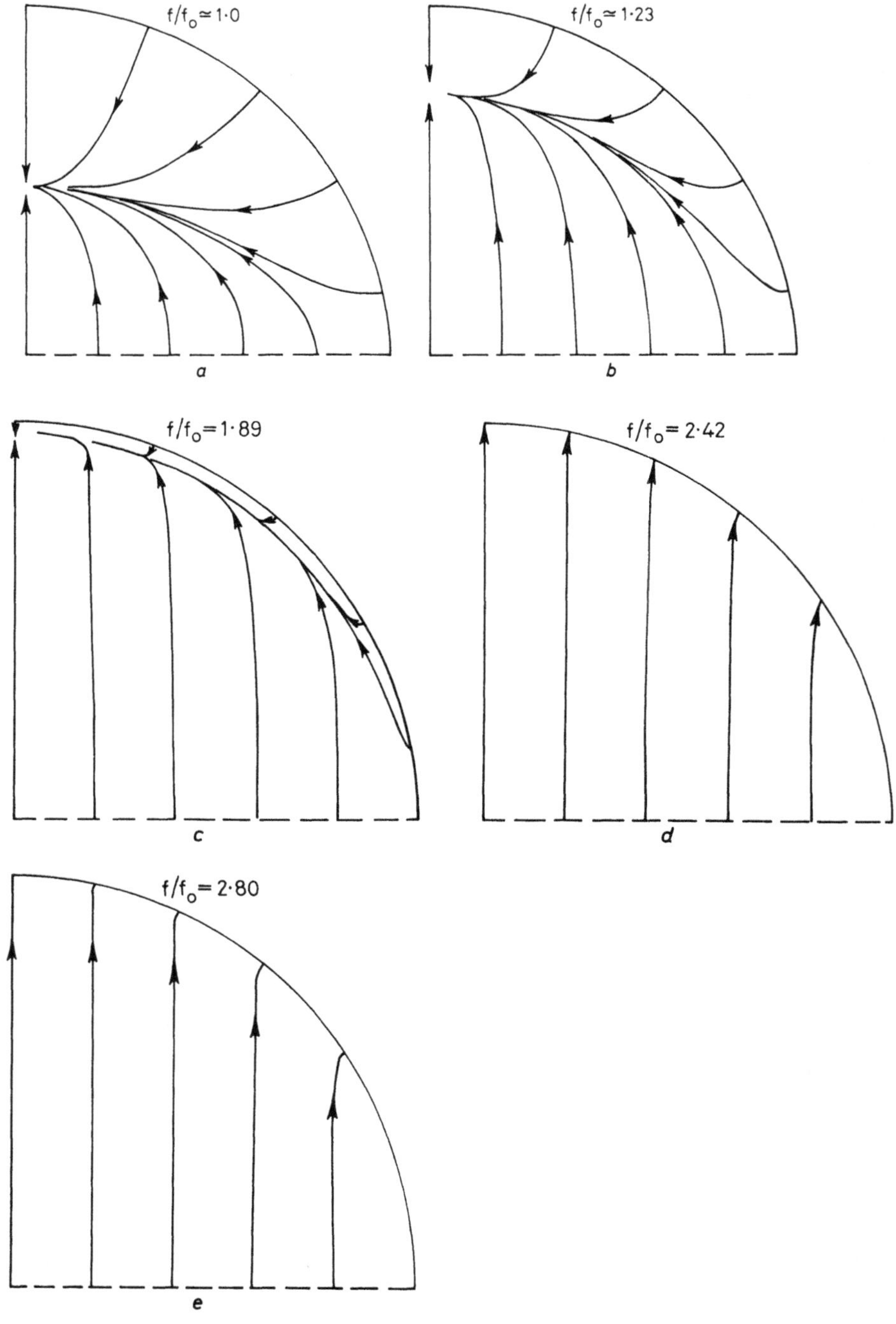

Fig. 3.13 *Transverse electric-field patterns of HE_{11} mode at various frequencies above cut-off*
a $f/f_c = 1{\cdot}0$ *b* $f/f_c = 1{\cdot}23$ *c* $f/f_c = 1{\cdot}89$ *d* $f/f_c = 2{\cdot}42$ *e* $f/f_c = 2{\cdot}80$

We know from our prior discussion that space-harmonics are required for an accurate representation of the fields in a periodic waveguide. The space-harmonic components of the fields are confined near the boundary $r = r_1$ where they modify the field matching conditions between the interior of the waveguide and the slot region. Chong [23] has made a detailed study of these features and below we present just a few of his results. First, in Figs. 3.12(a) and (b) we show how the amplitudes of the space-harmonics of the longitudinal electric and magnetic field vary with the order N of the harmonic. The oscillatory nature is evident together with the slow convergence. This behaviour explains the small change in the propagation coefficient between $K = 1$ and $K = 11$. In passing, we note that field matching demands the same number of slot modes as harmonics in the interior, thus we always chose $K = L$ in computations involving the space-harmonic model.

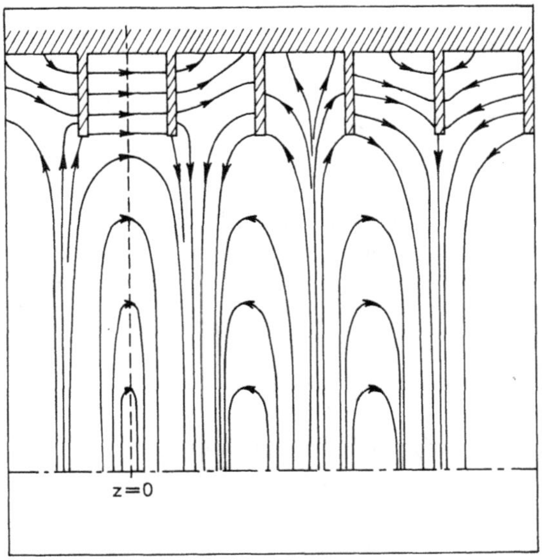

Fig. 3.14 *Longitudinal electric field pattern of HE_{11} mode: $r_1/r_0 = 0{\cdot}8$, $r_1 = 40$ mm, $\lambda_g = 30$ mm at 10·389 GHz*

Fig. 3.13(a)–(e) shows the transverse electric field pattern of the HE_{11} mode as the frequency f is increased above the cut-off value f_c. As indicated before, for these parameters, at cut-off the field of the HE_{11} mode exhibits pure TM_{11} properties but as frequency increases the mode changes becoming nearly a pure linearly polarised wave well above cut-off.

To complete our discussion Fig. 3.14 shows the electric field in a longitudinal section of the waveguide at an instant of time while Fig. 3.15 shows corresponding currents on a developed section of the corrugated wall. Here it must be said that discontinuities in current paths do arise because of the limitations imposed by inaccuracies in field matching, a consequence of limited computer resources.

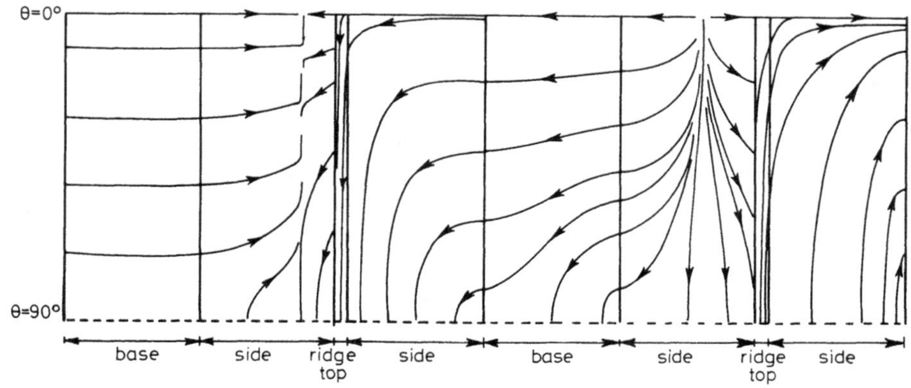

Fig. 3.15 *Current on the surface of the corrugations for HE_{11} mode (slots have been 'opened out' to display the currents). parameters as Fig. 3.14*

3.4 Attenuation in corrugated waveguides

Introduction

Although the attenuation of waveguides and horns used as a feed is so low as to be a relatively unimportant factor in design at present, the same does not necessarily hold for waveguides used as feeders, especially if the antenna is mounted remote from the transmitter. Furthermore both feed and feeder attenuation contributes directly to the system noise-temperature in satellite receiving terminals and this consitutes another reason for interest. Finally, with a progressive move towards higher microwave frequencies for telecommunications, attenuation will become an increasingly significant feature in design in the future.

A very detailed study of the attenuation of corrugated waveguides was made during the early 1970s by the authors and their co-workers at Queen Mary College, following a discovery made by one of them in 1969. Clarricoats observed that the HE_{11} mode of a corrugated waveguide had a lower value than that of the TE_{01} mode in a smooth-wall waveguide of comparable size. It will be recalled that at this time considerable commercial exploitation of the TE_{01} mode long-haul waveguide was expected so the discovery attracted some interest.

In simple terms the explanation for the phenomena is easy; the HE_{11} mode in a corrugated circular waveguide operating near the balanced-hybrid condition has only an H_z component of magnetic field tangential to the wall surface, i.e. $H_\phi \approx 0$. In common with the TE_{01} mode, this H_z component is very small compared to the transverse components of electric and magnetic field which support the power flowing through the waveguide and furthermore, it decreases with increasing normalised waveguide radius. Since attenuation is proportional to the strength of the field tangential to the metal surface the attenuation is correspondingly low. All fields in the slot region of the corrugated waveguide are proportional to the strength of the H_z field at the wall so, in spite of the increased wall area compared to a smooth wall waveguide, the attenuation is much lower. The fact that the HE_{11} mode has lower attenuation than the TE_{01} mode in a comparable size of waveguide can be explained by the presence of two loops of magnetic field for the TE_{01} mode compared to only one for the HE_{11} mode. The H_z field in the HE_{11} mode is thus lower for a given power flow. Compared to the dominant TE_{11} mode of smooth wall waveguide, the HE_{11} mode in a corrugated waveguide has an attenuation which is significantly lower, typically by one or two orders of magnitude.

Evaluation

The evaluation of the attenuation coefficient of a corrugated waveguide is presented below using the SI model. The derivation is straightforward and physical insight can be obtained from SI results. However, in contrast with the radiation characteristics of a corrugated waveguide, inclusion of space harmonics is vital to an accurate result for attenuation. The space harmonic formulation has been developed by us and is given in Reference 39. In this section our numerical results will draw on that theory.

We begin by deriving an expression for the Poynting vector S_z for $m = 1$ modes. In general,

$$S_z = E_x H_y - H_x E_y \tag{3.63}$$

For the $m = 1$ modes, we have on substitution from eqns. 3.59–3.62,

$$S_z = \frac{a_1^2}{4} y_0 \left(\frac{x'}{x}\right)^2 [\bar{\beta}(1+\bar{\Lambda}^2)\{J_0^2(x)+J_2^2(x)\}+\bar{\Lambda}(1+\bar{\beta}^2)$$
$$\{J_0^2(x)-J_2^2(x)\}-2\bar{\beta}(1-\bar{\Lambda}^2)J_0(x)J_2(x)\cos 2\phi\} \tag{3.64}$$

Eqn. 3.64 shows incidentially that when $\bar{\Lambda} = \pm 1$ the energy distribution is circularly symmetric. Further if $\bar{\Lambda} \simeq \bar{\beta} \simeq 1$ we have

$$S_z = a_1^2 y_0 \left(\frac{x'}{x}\right)^2 J_0^2(x) \tag{3.65}$$

To obtain the total power flowing through the region $r < r_1$ P_T, we must integrate S_z over $0 < r < r_1$. Then

$$P_T = \frac{a_1^2}{k^2}\pi\left(\frac{x'}{x_1}\right)^4 y_0 J_1^2(x_1)[\bar{\beta}(1+\bar{\Lambda}^2)\{\tfrac{1}{2}F_1^2(x_1)+F_1(x_1)$$
$$+\tfrac{1}{2}(x_1^2-1)\}+\bar{\Lambda}(1+\bar{\beta}^2] \tag{3.66}$$

In the same approximation as above, noting $F_1(x_1) \simeq -1$,

$$P_T = \frac{a_1^2}{k^2}\pi(x_1')^4 y_0 \left\{\frac{J_1^2(x_1)}{(x_1)^2}\right\} \tag{3.67}$$

The above results are for circularly polarised waves; for linearly polarised waves the power flow is halved.

If space harmonics and higher-order slot modes are neglected, a simple expression for the attenuation coefficient α in a corrugated circular waveguide can be obtained following lengthy algebraic manipulations.

$$\alpha = \frac{P_L}{2 P_T b} \tag{3.68}$$

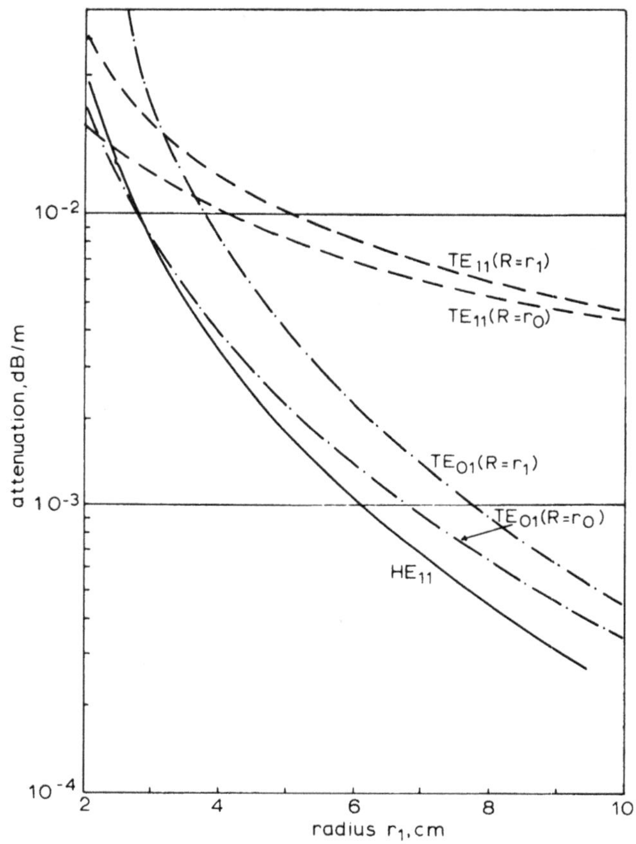

Fig. 3.16 *Comparison of HE_{11} mode attenuation as a function of radius in brass corrugated waveguide and TE_{11} mode and TE_{01} mode in brass smooth-wall waveguide. Frequency = 9 GHz. $s/\lambda = 0{\cdot}25$. $b/\lambda = 0{\cdot}3$. $t/b = 0{\cdot}1$*

where

$$P_L = P_{L_1} + P_{L_2} \tag{3.69}$$

and $P_{L_{1,2}}$ is the power lost respectively on the side-walls and base of the corrugations. b is the slot width and the ridge thickness is assumed negligible.

$$P_{L_1} = a_1^2 \frac{2\pi}{k^2} R_S\, y_0^2 J_1^2(x_1)[\tfrac{1}{2}(1 - x_1'^2) + \frac{2}{\{\pi S_1^D(x_1', x_0')\}^2} - S_1(x_1', x_0') - \tfrac{1}{2} S_1^2(x_1', x_0')] \tag{3.70}$$

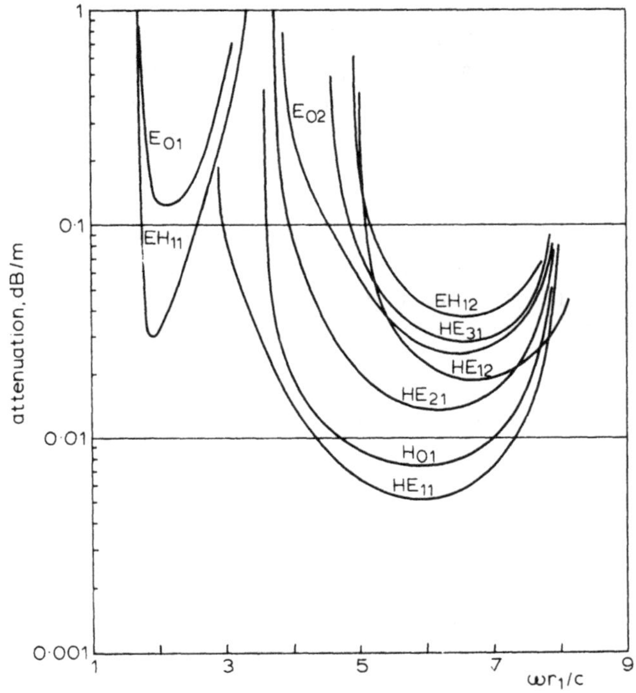

Fig. 3.17 *Attenuation characteristics of corrugated circular waveguide. Conductivity = 5·8 × 10^7 S/m, r_1 = 25 mm, s = 10 mm, b = 10 mm, t/b = 0·2*

$$P_{L_2} = \frac{4R_S b y_0^2 J_1^2(x_1) a_1^2}{k^2 r_0\, [S_1^D(x_1', x_0')]^2} \tag{3.71}$$

R_S is the resistivity of the metal from which the waveguide is constructed. The functions $S_1^D(x, y)$ and $S_1(x, y)$ are defined in eqns. 3.14–3.15.

Under balanced hybrid conditions eqns. 3.70 and 3.71 simplify so that when combined with eqn. 3.67 for P_T, the expression for α becomes, when $r_1/r_0 \simeq 1$

$$\alpha = \frac{1}{2}\left(\frac{R_S}{z_0}\right)\frac{x_1^2 k}{x_1'^4}\left[\frac{1}{kb}\left(1 + \frac{x_1'\pi}{2}\right) + x_1'\right] \tag{3.72}$$

when $x_1' \gg 1$

$$\alpha \simeq \frac{1}{2}\left(\frac{R_S}{z_0}\right)\frac{x_1^2 k}{(x_1')^3}\left[\frac{\pi}{2(kb)} + 1\right] \tag{3.73}$$

Eqn. 3.73 shows that the attenuation at the balanced-hybrid condition falls as $k/(x_1')^3$ which implies a dependence upon r_1 of r_1^{-3} and upon frequency f of f^{-2}. The resistivity depends upon f as $f^{1/2}$ leading to an overall frequency dependence of $f^{-3/2}$. These are the same characteristics as are exhibited by the TE_{01} mode in a smooth wall waveguide. As mentioned previously, however, the results of exact calculations shown in Fig. 3.16 show that the HE_{11} mode in a corrugated waveguide has lower attenuation than the TE_{01} mode in a waveguide of comparable size. Fig. 3.17 shows the attenuation for higher order modes in addition to the HE_{11} mode.

An extensive programme of measurements made by Clarricoats, Olver and Chong [39], also by Parini [129] have supported the theoretical results presented above. The interested reader is referred to these for further details.

3.5 Radiation from corrugated waveguides

3.5.1 Introduction

Two methods have been used to determine the radiation pattern of cylindrical corrugated waveguides: the Kirchhoff–Huygen (KH) method and the Fourier Transform (FT) method. For a complete description of the radiated field, both methods require a description of transverse fields over the waveguide aperture, over the waveguide flange and over the remainder of the aperture plane. The KH method requires both the transverse electric and transverse magnetic field to be specified, whereas the FT method only requires the transverse electric field. Since the transverse electric field vanishes over the assumed perfectly conducting flange and is very small beyond, the FT method yields accurate results when the field beyond the radiating aperture is assumed to be zero. This assumption is equivalent to an infinite flange. To obtain comparable accuracy with the KH method would require a knowledge of the currents over the flange in order to describe the transverse magnetic field there and this information is not readily available.

When the aperture is large compared to a wavelength the currents over the flange are very small and then the two methods yield very similar patterns. To improve on the approximate FT method a correction for the finite size of the flange can be made using the GTD method.

From the above it is clear that the FT method offers both accuracy and simplicity and it is used in the treatment of radiation from the corrugated waveguide described below. Readers interested in details of the KH method are referred to Section 4.5 where results for conical horns are given.

3.5.2 Radiation from a circular aperture

If, as in Fig. 3.18(a), the Cartesian components of electric field E_x, E_y are specified over the aperture $0 < r < r_1$, $0 < \phi' < 2\pi$, the radiated components of electric field at R, θ, ϕ are given by

$$E_{\substack{x\\y}p} = A\int_0^{2\pi}\int_0^{r_1} E_{\substack{x\\y}} \exp\left(jkr\sin\phi\cos\left(\phi-\phi'\right)\right) r\,dr\,d\phi' \tag{3.73}$$

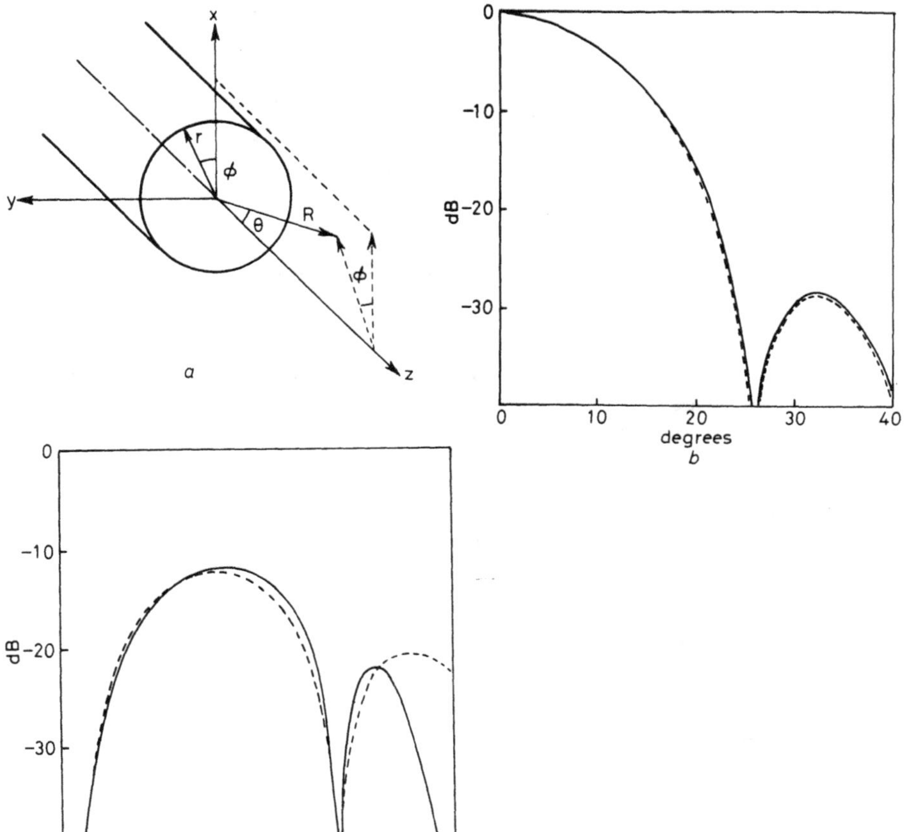

Fig. 3.18 *a Geometry for radiation from corrugated waveguides*
b Copolar radiation patterns for a corrugated waveguide of aperture diameter = 4λ
——— eqn. 3.84
– – – exact solution using space harmonics
c Crosspolar radiation patterns corresponding to Fig. 3.18 (b)
——— eqn. 3.85,
– – – exact solution using space harmonics

where

$$A = \frac{jk}{2\pi R} e^{-jkR} \cos\theta$$

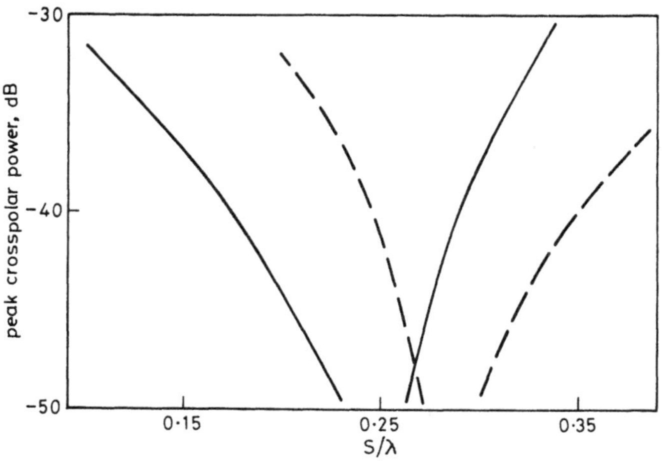

d

d Peak crosspolar power as a function of frequency corresponding to Fig. 3.18 (c) (ridge width/slot width = 0·2 for space harmonic model)

Initially we neglect the effect of space harmonics and insert E_x and E_y from eqns. 3.59–3.60 into eqn. 3.73 so as to obtain for $n = 1$ modes

$$E_{xp} = AB\{(\bar{\Lambda} + \bar{\beta})N_0 + (\bar{\Lambda} - \bar{\beta})N_2 \cos 2\phi\} \tag{3.74}$$

$$E_{yp} = AB\{(\bar{\Lambda} - \bar{\beta})N_2 \sin 2\phi\} \tag{3.75}$$

where

$$B = -ja_1 \frac{k}{K} \pi$$

$$N_k = \frac{r_1^2}{v^2 - x_1^2} \{vJ_k(x_1) J_1(v) - x_1 J_k(v) J_1(x_1)\} \quad k = 0, 2$$

$$x_1 = Kr_1$$

$$v = kr_1 \sin\theta$$

To obtain the copolar and crosspolar fields (the latter in the third definition given by Ludwig, see Section 5.2) we use

$$E_{co} = E_{xp}\left(\cos^2\frac{\theta}{2} - \sin^2\frac{\theta}{2}\cos 2\phi\right) - E_{yp}\sin^2\frac{\theta}{2}\sin 2\phi \tag{3.76}$$

$$E_{cr} = -E_{xp}\sin^2\frac{\theta}{2}\sin 2\phi + E_{yp}\left(\cos^2\frac{\theta}{2} + \sin^2\frac{\theta}{2}\cos 2\phi\right) \tag{3.77}$$

On substitution for E_{xp} and E_{yp}

$$E_{co} = AB\left\{[(\bar{\Lambda}+\bar{\beta})N_0 + (\bar{\Lambda}-\bar{\beta})N_2\cos 2\phi]\left(\cos^2\frac{\theta}{2} - \sin^2\frac{\theta}{2}\cos 2\phi\right) - (\bar{\Lambda}-\bar{\beta})N_2\sin^2\frac{\theta}{2}\cos 2\phi\sin 2\phi\right\} \tag{3.78}$$

In the principal planes where $\phi = n\dfrac{\pi}{2}$ and $n = 0, 1, 2, 3$

$$E_{co} = AB\left\{[(\bar{\Lambda}+\bar{\beta})N_0 \pm (\bar{\Lambda}-\bar{\beta})N_2]\begin{bmatrix}\cos\theta\\ 1\end{bmatrix}\right\} n = \begin{bmatrix}0, & 2\\ 1, & 3\end{bmatrix} \tag{3.79}$$

In the 45° planes where $\phi = n\dfrac{\pi}{4}$ and $n = 1, 3, 5, 7$

$$E_{co} = AB\left\{(\bar{\Lambda}+\bar{\beta})N_0\cos^2\frac{\theta}{2}\right\} \tag{3.80}$$

The crosspolar field vanishes in the principal planes while in the 45° planes, where it attains a maximum value off-axis,

$$E_{cr} = AB\left\{-(\bar{\Lambda}+\bar{\beta})N_0\sin^2\frac{\theta}{2} + (\bar{\Lambda}-\bar{\beta})N_2\cos^2\frac{\theta}{2}\right\} \tag{3.81}$$

The ratio $\dfrac{E_{cr}}{E_{co}(\theta = 0)} \equiv C$ is given by

$$C = -\sin^2\frac{\theta}{2} + \frac{(\bar{\Lambda}-\bar{\beta})}{(\bar{\Lambda}+\bar{\beta})}\frac{N_2}{N_{0\theta=0}}\cos^2\frac{\theta}{2} \tag{3.82}$$

For the HE_{11} mode in a large aperture waveguide $\bar{\beta} \approx 1, x_1 \approx 2.405$, and following Dragone [56], we can show that the maximum in the cross polar pattern occurs when $v = 3.67$, then if $\sin^2\dfrac{\theta}{2} \ll 1$,

$$C = (0.26)\left[\frac{\bar{\Lambda}-1}{\bar{\Lambda}+1}\right] \tag{3.83}$$

Recalling that $v = kr_1\sin\theta$, the above expression is valid when kr_1 is greater than about 8.

Let us now examine the patterns predicted by the above equations. Fig. 3.18(b) shows for $2r_1/\lambda = 4$, the normalised copolar power pattern

$$P_{co}(\theta) = \frac{N_0^2(\theta, x_1)}{N_0^2(0, x_1)} = \left[\frac{x_1^2 J_0(v)}{x_1^2 - v^2}\right]^2 \tag{3.84}$$

obtained by referring the radiated power to that in the boresight direction and

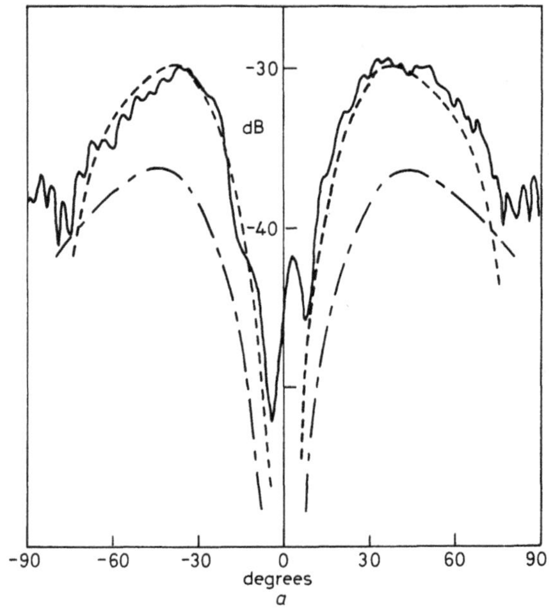

a

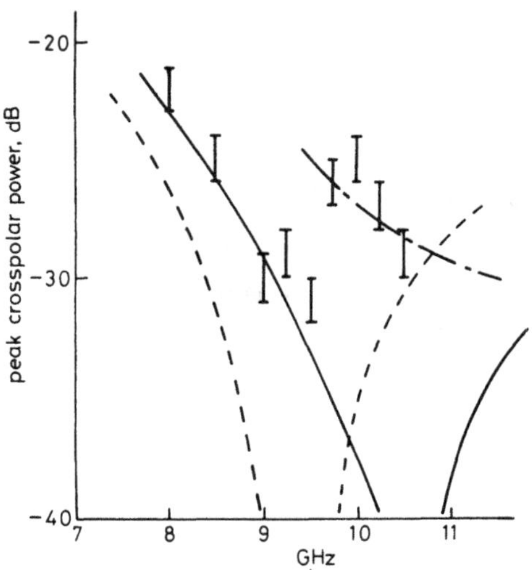

b

Fig. 3.19 *a Crosspolar radiation patterns of open-ended corrugated waveguide*
—— measured – – – computed with Fourier Transform (FT) theory —·— computed with Kirchhoff–Huygen (KH) theory Geometric parameters as Fig. 3.17.
b Peak crosspolar power as a function of frequency for waveguide of Fig. 3.19(a)
—— $K = L = 1$ – – – $K = L = 0$ —·— HE_{12} mode crosspolarisation. Vertical bars indicate measured values

assuming that $x_1 = 2{\cdot}405$, Fig. 3.19(c) shows for the angles where $\sin^2 \dfrac{\theta}{2} \ll 1$

$$\frac{P_{cr}}{\left(\dfrac{\bar{\Lambda}-\bar{\beta}}{\bar{\Lambda}+\bar{\beta}}\right)^2} = \frac{N_2^2(\theta, x_1)}{N_0^2(0, x_1)} = \left(\frac{x_1^2}{x_1^2 - v^2}\right)^2 [vJ_2(x_1)J_1(v) - x_1 J_2(v) J_1(x_1)] \tag{3.85}$$

In both cases the results are compared with predictions based on FT theory applied to the aperture fields including $N = \pm 1$ space harmonics in addition to the fundamental. The agreement is excellent for the coplar pattern and satisfactory for the cross-polar pattern. However, the absolute value of P_{cr} requires an accurate estimate of the factor $[(\bar{\Lambda} - \bar{\beta})/(\bar{\Lambda} + \bar{\beta})]^2$. For other than small aperture waveguides we can obtain $\bar{\beta}$ with considerable accuracy from the expression

$$\bar{\beta} = 1 - \tfrac{1}{2} \frac{x_1^2}{(kr_1)^2} \tag{3.86}$$

with $x_1 = 2{\cdot}405$.

$\bar{\Lambda}$ depends much more critically on the corrugation parameters, as discussed in Chapter 5, and to obtain a useful estimate of the crosspolar level the value of $\bar{\Lambda}$ must be obtained from a solution of the characteristic equation including the ± 1 space harmonics in addition to the fundamental, as Fig. 3.18(d) shows.

In practice, feeds whose circular aperture is preceded by a section of cylindrical corrugated waveguide are of two kinds; those of small diameter, where to obtain accurate predictions eqns. 3.78 and 3.81 must be used without approximations and a correction made for the finite size of the flange, and feeds of large diameter forming part of a so-called profiled horn. In the first category, Fig. 3.19(a) shows results for a corrugated waveguide of 5 cm internal radius measured at wavelengths around 3 cm. The patterns compare well with theory, provided space harmonics are included, and the graph of maximum crosspolar intensity as a function of frequency, Fig. 3.19(b), also agrees well, although the effect of higher mode excitation is in evidence at higher frequencies.

For large diameter corrugated feeds the intrinsic radiation from the HE_{11} mode can be predicted with accuracy using the approximate eqns. of 3.84 and 3.85. To achieve these aperture sizes of greater than about 5λ in a reasonably short length, the aperture must be connected to the input waveguide by means of a profiled section and in this section higher mode conversion can occur unless the design is optimum. Thus, in general, the patterns will be disturbed by the presence of higher modes in the radiating aperture. These must be included for accurate prediction, but the main design objective is to minimise their excitation. Further discussion of the profiled horn will be deferred until mode excitation has been discussed in Section 4.4. Additional design information can be found in Chapter 5, Section 5.8.7.

A large aperture can also be achieved with a narrow flare-angle conical horn. Results for such horns, where higher modes are minimal, approximate well with those of large aperture corrugated waveguides. The agreement between experimental and theoretical results is very good, but to obtain precise correlation account must be taken of the spherical wavefront. Discussion of this well be taken up again in Chapter 4, see Figs. 4.21 to 4.23.

3.6 Junctions in corrugated waveguides

The most common means to excite the dominant HE_{11} mode of a circular cross-section corrugated waveguide is from a circular waveguide with a uniform wall supporting the TE_{11} mode. The problem was first analysed by Clarricoats and Saha [29] with later studies by Cooper [49], Dragone [56], James [74, 75, 77] and Kuhn [171], among others. The configuration is shown in the inset to Fig. 3.20.

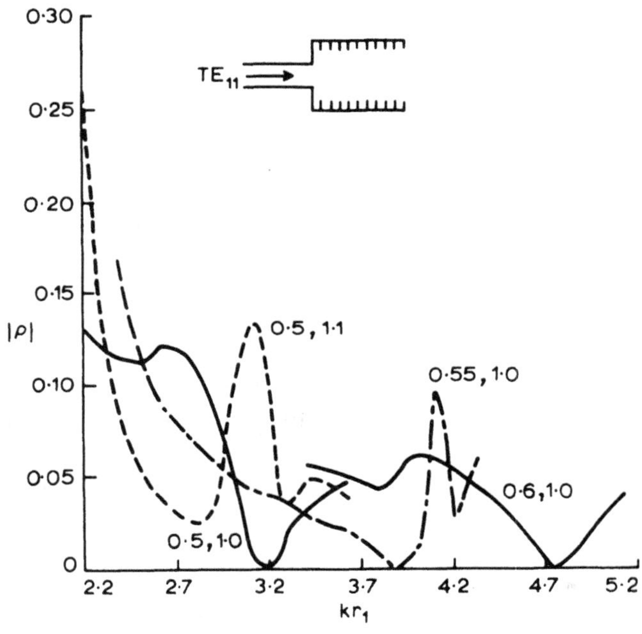

Fig. 3.20 *Reflection coefficient of a TE_{11} mode incident on a smooth-wall to corrugated waveguide junction. Parameters r_1/r_0 and r_1/r_h, r_h = radius of smooth wall waveguide*

To determine the reflection and transmission coefficients at the junction, we expand the electric and magnetic fields in the plane of the aperture in terms of normal modes of each waveguide. An outline of the modal expansion method now follows. If $\mathbf{e}_{t_m}$, $\mathbf{h}_{t_m}$ and $\mathbf{e}_{t_M} \mathbf{h}_{t_M}$ are, respectively, the transverse electric and magnetic fields of the mth and Mth mode in the two waveguides,

$$(1+\rho)\,\mathbf{e}_{t_1} + \sum_{m=2}^{\infty} a_m \mathbf{e}_{t_m} = \sum_{M=1}^{\infty} A_M \mathbf{e}_{t_M} \tag{3.87}$$

$$(1-\rho)\,\mathbf{h}_{t_1} - \sum_{m=2}^{\infty} a_m\,\mathbf{h}_{t_m} = \sum_{M=1}^{\infty} A_M\,\mathbf{h}_{t_M} \tag{3.88}$$

By applying the orthogonality condition

$$\int_S \mathbf{e}_{t_n} \times \mathbf{h}^*_{t_m}\,\mathbf{dS} = 0 \qquad m \neq n \tag{3.89}$$

and the corresponding equation for modes in the other waveguide, a doubly infinite matrix of equations is generated. By taking a finite number of terms, i.e. a

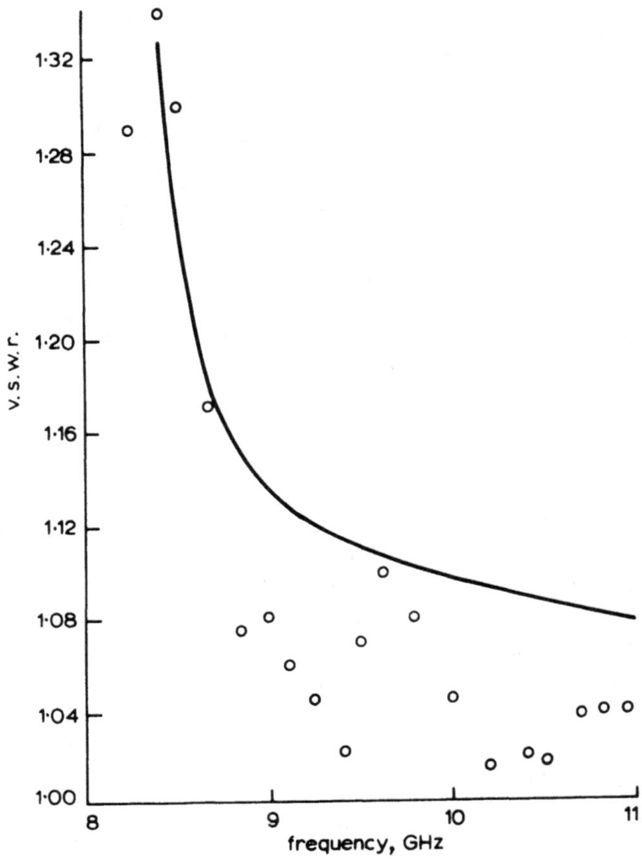

Fig. 3.21 *Measured VSWR of a 12 degree semi-flare angle corrugated horn and theoretical curve for* $r_1/r_0 = 0.22$ *and* $r_1/r_h = 1.1$

finite number of modes in each waveguide, the equations can be solved. In general the coefficients a_m and a_M are complex and if p modes are taken in each waveguide a $4p \times 4p$ matrix equation has to be solved.

In a rigorous solution, fields in the corrugated waveguide must be obtained using the space-harmonic formulation discussed in Section 3.2, but to a good approximation, in most cases they can be neglected and a surface impedance model used instead.

Fig. 3.20 shows theoretical values of TE_{11} mode reflection coefficient obtained

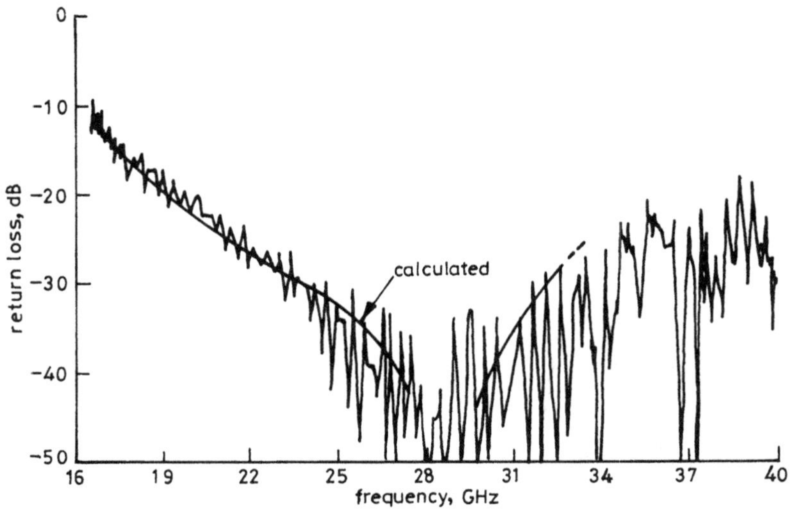

Fig. 3.22 *Return loss of a corrugated horn with semi-flare angle of 4 degrees and aperture radius 3·14 cm [56]*

by Clarricoats and Saha [29] in a study in which the parameters r_1/r_0 and r_1/r_h were varied r_h being the radius of the uniform circular waveguide.

A comparison between experiment and theory is shown in Fig. 3.21 where the agreement lies within experimental accuracy. It is to be noted that the reflection coefficient falls to zero at those frequencies where in the surface impedance model of the corrugated waveguide, the wall impedance is zero, i.e. the corrugation depth is approximately $\lambda/2$.

In an early study of hybrid mode waveguides made by Clarricoats, an approximate value for ρ was shown to be given by

$$|\rho| = \left|\frac{\beta_1 - \beta_2}{\beta_1 + \beta_2}\right| \tag{3.90}$$

where β_1 and β_2 are the propagation coefficients of the modes of the two waveguides. This result was established independently by Dragone [56] more recently and Fig. 3.22, due to him, satisfactorily compares eqn. 3.90 with experimental results.

James [74, 75, 77] has approached the discontinuity problem from a different standpoint. He treats the corrugated waveguide as a cascade of uniform waveguides of very short length with radii corresponding to the inside of the corrugated waveguide and the base of the corrugation. The approach has the virtue of allowing one to treat transitions in the throat region of a corrugated waveguide where the corrugation depth changes rapidly with distance along the waveguide. Fig. 3.23 shows James's prototype configuration and theoretical return loss for different types of transition. We notice that the transition yielding lowest return loss is one in which depth is varied from $\lambda/2$ at the throat entrance towards $\lambda/4$ at the exit.

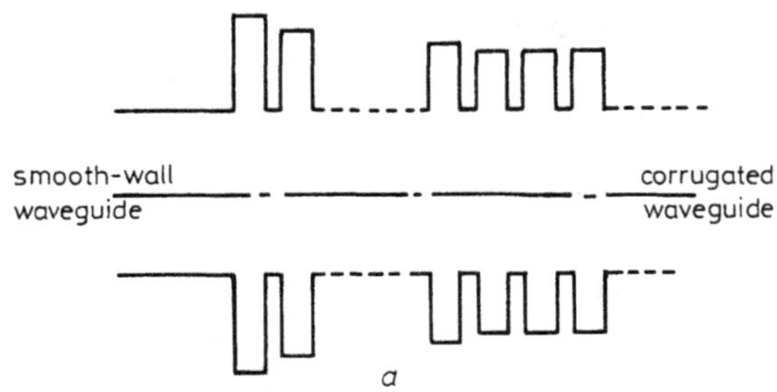

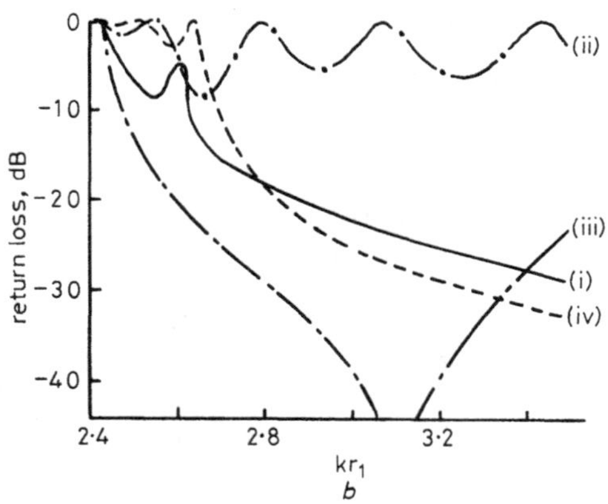

Fig. 3.23 *a Throat region of corrugated horn showing junction between smooth-walled waveguides of differing radius.*
b Theoretical return loss of seven slot corrugated mode converter
(i) Return-loss without convertor
(ii) Convertor with slot depth increasing from zero
(iii) Convertor with decreasing slot depth, beginning with $\lambda/2$ at the input (f_0 is the centre frequency where $kr_1 = 2{\cdot}9$)
(iv) Convertor with constant slot depth but varying t/b

A convertor with constant slot depth of $\lambda/4$ is seen to be the least favourable of those studied. Kuhn [171] has recently solved the discontinuity problem using a scattering matrix formulation. The design of transitions is discussed again in Chapter 5.

3.7 Multimode corrugated waveguides

The feeds considered above are designed to support one mode of propagation and a higher-order mode is present only as an undesirable feature. This situation contrasts with the multimode feed where the presence of the higher-order mode is an essential and desirable aspect of the design. Multimode feeds can be designed using smooth-wall waveguide but in general their features are similar to those of multimode

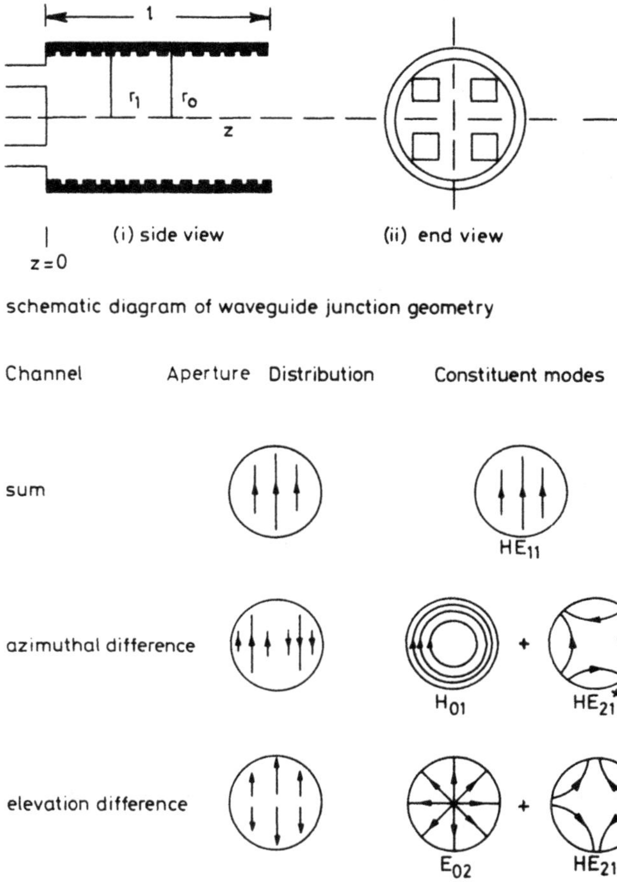

Fig. 3.24 *Corrugated monopulse feed and monopulse mode patterns*

corrugated feeds, with one exception. The exception is the so-called Potter horn [188], where a hybrid field is synthesised by the excitation of a TE_{11} and TM_{11} mode in correct amplitude and phase. Such a feed emulates the HE_{11} mode of a corrugated feed, but over a limited bandwidth. Returning to multimode corrugated feeds *per se*, two principal applications exist: in one, higher modes are present to improve efficiency; in another, they provide tracking information. An example of the first kind is the feed combining the HE_{11} and HE_{12} modes which was developed as a prime-focus feed for the Parkes radiotelescope by Vu and Vu [157]. Another example in this class, is the feed combining HE_{11} and HE_{21} modes proposed by Rudge and Adatia [189, 190, 132] to synthesise the focal field of an offset reflector antenna. In the second class lies the multimode feed shown in Fig. 3.24 together with corresponding modal fields. The HE_{11} mode provides the signal channel while combinations of orthogonal HE_{21} modes with, respectively, H_{01} and E_{02} modes provide azimuth and elevation outputs. A feed of this kind has the advantage over a pure-mode equivalent in that it affords polarisation purity and the concomitant opportunity to extract polarisation information in sum and difference channels.

The feed is operated at a frequency such that the normalised propagation coefficient of all three of the tracking modes are nearly degenerate and lie close to the condition $\bar{\beta} = \bar{\Lambda}$.

General expressions for the fields are as follows:

HE_{11} mode (sum channel)

$$E_x = -\mathrm{j}\frac{a_{11}}{2}\frac{k}{K}\{(\bar{\beta}+\bar{\Lambda})J_0(Kr) + (\bar{\Lambda}-\bar{\beta})J_2(Kr)\cos 2\phi\} \tag{3.91}$$

$$E_y = -\mathrm{j}\frac{a_{11}}{2}\frac{k}{K}\{(\bar{\Lambda}-\bar{\beta})J_2(Kr)\sin 2\phi\} \tag{3.92}$$

The form of the equation E_y shows that the aperture field becomes perfectly linearly polarised when $\bar{\Lambda} = \bar{\beta}$. However, recalling eqn. 2.83, other than for very large apertures, the condition for zero cross polarisation of the radiated field occurs at a frequency slightly above that at which $\bar{\Lambda} = 1$. This latter condition is referred to as balanced hybrid, and when it occurs the corrugation depth of a waveguide with $kr_1 \gg 1$ is approximately $\lambda/4$, as explained in section 3.2.

H_{01} mode (contributor to the azimuth-difference channel)

$$E_x = \mathrm{j}y_0 b_{01}\frac{k}{K}J_1(Kr)\sin\phi \tag{3.93}$$

$$E_y = -\mathrm{j}y_0 b_{01}\frac{k}{K}J_1(Kr)\cos\phi \tag{3.94}$$

E_{02} mode (contributor to the elevation-difference channel)

$$E_x = \mathrm{j}a_{02}\frac{k}{K}\bar{\beta}J_1(Kr)\cos\phi \tag{3.96}$$

$$E_y = \mathrm{j}a_{02}\frac{k}{K}\bar{\beta}J_1(Kr)\sin\phi \tag{3.96}$$

HE_{21} (contributor to the elevation-difference channel)

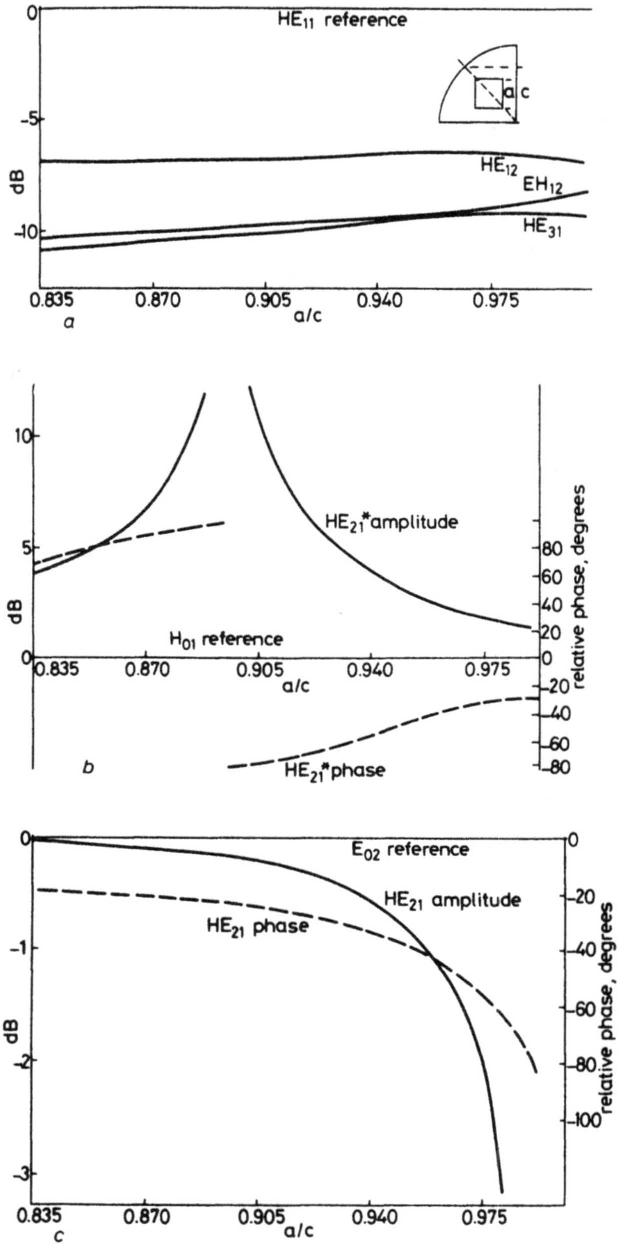

Fig. 3.25 *Mode excitation levels as a function of a/c for square waveguide with waveguide centres located at $r_1/2$*

a Excitation levels of modes in sum channel operation

b Excitation levels of modes in azimuthal difference channel operation

c Excitation levels of modes in elevation difference channel operation

HE_{21*} (contributor to the azimuth-difference channel – asterisk denotes orthogonal polarisation)

$$E_{\substack{x*\\y}} = -\mathrm{j}\,\frac{a_{21}}{2}\,\frac{k}{K}\,\{(\bar{\beta}+\bar{\Lambda})\,J_1(Kr)\cos\phi + (\bar{\Lambda}-\bar{\beta})\,J_3(Kr)\cos 3\phi\} \quad (3.97)$$

$$E_{\substack{y*\\x}} = \pm\mathrm{j}\,\frac{a_{21}}{2}\,\frac{k}{K}\,\{(\bar{\beta}+\bar{\Lambda})\,J_1(Kr)\sin\phi + (\bar{\Lambda}-\bar{\beta})\,J_3(Kr)\sin 3\phi\} \quad (3.98)$$

In contrast with the HE_{11} mode, the equation above shows that for the HE_{21} mode, even if the term in $(\bar{\Lambda}-\bar{\beta})$ is negligible, the field has equal copolar and crosspolar components. But, assuming mode degeneracy, a judicious choice of the ratio of coefficients a_{02} and a_{21} ensures, provided the fields are combined in-phase, that the crosspolar component of the total field vanishes. Then the linearly-polarised aperture field under nearly balanced-hybrid conditions generates an almost linear-polarised far field. Similar arguments apply for the H_{01} and HE_{21*} modes.

In summary, the contributions to E_y from the H_{01}, E_{02} and two types of HE_{21} modes cancel when the terms in $(\bar{\Lambda}-\bar{\beta})$ are neglected, then the aperture fields of the difference channels have the form.

$$E_x = C_1\,J_1(Kr)\cos\phi\ \text{(elevation)} \quad (3.99)$$

$$E_x = C_2\,J_1(Kr)\sin\phi\ \text{(azimuth)} \quad (3.100)$$

As noted previously, any departure from the correct amplitude or phase of the combination will introduce crosspolarisation, as also will a significant departure from the condition that $(\bar{\Lambda}-\bar{\beta})$ is negligible for the HE_{21} mode.

Clarricoats and Elliot [25] have made a detailed analysis of a feed of the kind shown in Fig. 3.24 and have examined how the excitation coefficients depend on the size and location of the square input waveguides. Typical results are shown in Fig. 3.25. From their parametric study they conclude that a cluster of square waveguides close to the centre of the feed is best to achieve appropriate excitation of all the modes, although they also demonstrate the need for a mode-selective phase-shifter in order to produce the correct phase characteristics in the feed aperture.

A similar multimode feed which just uses the HE_{11}, HE_{21} and E_{02} modes has been developed by Watson, Dang and Ghosh [161]. In this feed the azimuth and elevation channels are extracted from a sum and difference of the latter mode pair. Their feed system was developed for use in a spacecraft where accurate alignment of the antenna with respect to a point on earth, is an essential requirement for efficient transmission.

Chapter 4

Propagation and radiation characteristics of conical corrugated waveguides

4.1 Introduction

Many corrugated feeds used with reflector antennas take the form of conical horns or have a conical region even though the aperture region is cylindrical. For horns with small semi-flare angles, in the range 6° to 15°, the horn behaves as a cylindrical waveguide but a correction must be made for the spherical phase variation across the horn aperture. This procedure, which was explored by Parini and Olver [127], enables one to characterise the aperture field very accurately since space harmonics can be described in a cylindrical corrugated waveguide, as discussed in Chapter 3. Once we have the aperture field, the radiation field can be obtained using the Fourier Transform method described in Section 3.5.

An alternative method which works well over the range of flare angles from about 10° to 80° uses spherical modes to describe the field in the horn and in particular in the horn aperture. The radiation pattern can be obtained by integration over the spherical phase surface or by means of an expansion. Two expansion methods have been tried: the spherical wave expansion method, developed by Clarricoats, Saha and Olver [30, 43] and extended by Mahmoud and Clarricoats [95]; and the Laguerre–Gaussian method developed by Bitter and Aubry [17]. The latter method is useful when a corrugated feed forms part of a multiple reflector system designed using Gaussian beam approximations.

In this chapter we shall explore these methods, concentrating especially on the spherical wave expansion method because of its wide applicability. We shall also have need to explore mode conversion along the horn for in many cases the cross-polar radiation pattern will be dominated by an undesired higher-order mode rather than the dominant HE_{11} spherical hybrid mode.

4.2 Modal characteristics of corrugated conical horns

A conical corrugated horn is shown in Fig. 4.1(a). To a good approximation the field in the horn may be described in terms of spherical modes which are hybrid unless the field has zero azimuthal dependence. The boundary condition at the wall $\theta = \theta_1$ presents a problem since it is not possible to describe fields in the slot rigorously. However, one approximate method due to Jansen, Jeuken and Lambrechtse [80], which works quite adequately in the aperture region of the horn, treats the slot as if it were part of a conical waveguide, as shown in Fig. 4.1(b).

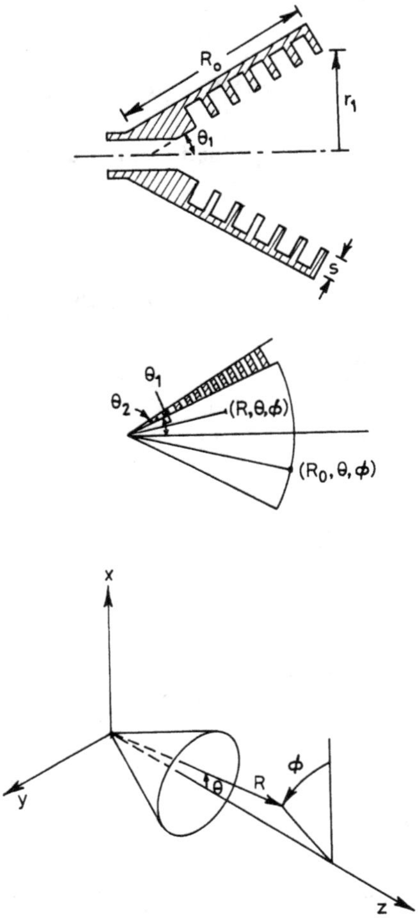

Fig. 4.1 *Conical corrugated horns*

We can then describe the boundary through an average slot admittance $Y = -(H_\phi/E_R)$ an expression for which is given below (eqn. 4.5). In spite of its simplicity the average surface impedance model has been found to work very well and especially for wide angle horns.

We shall now develop expressions for the fields in the horn using the spherical mode method.

Modal fields

As previously remarked the modes are, in general, hybrid and both electric and magnetic fields exist in the outward radial direction R of Fig. 4.1(a). These can be written for any particular mode, assuming a time dependence exp (jωt) as

$$E_R = A\,\frac{\nu(\nu+1)}{\mathrm{j}kR^2}\,\hat{H}_\nu^{(2)}(kR)P_\nu^m(\cos\theta)\cos m\phi \tag{4.1a}$$

$$y_0 H_R = A\,\bar{\Lambda}\,\frac{\nu(\nu+1)}{\mathrm{j}kR^2}\hat{H}_\nu^{(2)}(kR)P_\nu^m(\cos\theta)\sin m\phi \tag{4.1b}$$

$P_\nu^m(\cos\theta)$ is an associated Legendre function of order m and degree ν, while $H_\nu^{(2)}(kR)$ is a spherical Hankel function of order ν. $\bar{\Lambda}$ is the hybrid factor possessing a characteristic frequency dependent value for each mode. Knowing E_R and H_R, the transverse fields are obtained as

$$RE_\theta = -A\,\hat{H}_\nu^{(2)}(kR)\,[\bar{\beta}P_\nu'^m(\cos\theta) + m\bar{\Lambda}P_\nu^m(\cos\theta)/\sin\theta]\,\cos m\phi \tag{4.2a}$$

$$RE_\phi = A\hat{H}_\nu^{(2)}(kR)\,[\bar{\Lambda}P_\nu'^m(\cos\theta) + m\bar{\beta}P_\nu^m(\cos\theta)/\sin\theta]\,\sin m\phi \tag{4.2b}$$

$$y_0 RH_\theta = -A\hat{H}_\nu^{(2)}(kR)\,[\bar{\Lambda}\bar{\beta}P_\nu'^m(\cos\theta) + mP_\nu^m(\cos\theta)/\sin\theta]\,\sin m\phi \tag{4.2c}$$

$$y_0 RH_\phi = -A\hat{H}_\nu^{(2)}(kR)[P_\nu'^m(\cos\theta) + m\bar{\Lambda}\bar{\beta}P_\nu^m(\cos\theta)/\sin\theta]\,\cos m\phi \tag{4.2d}$$

where $\bar{\beta}$ is defined by

$$\bar{\beta}(R) = \mathrm{j}\hat{H}_\nu'^{(2)}(kR)/\hat{H}_\nu^{(2)}(kR) = \mathrm{j}\frac{\partial}{\partial kR}\{\ln[\hat{H}_\nu^{(2)}(kR)]\} \tag{4.3}$$

from which it follows that

$$\hat{H}_\nu^{(2)}(kR)/\hat{H}_\nu^{(2)}(kR_0) = \exp\left[-\mathrm{j}k\int_{R_0}^{R}\bar{\beta}(R)\,\mathrm{d}R\right] \tag{4.4}$$

$\bar{\beta}$ is recognised as a normalised complex propagation parameter. Various approximate formulae for $\bar{\beta}$ in various ranges of kR are summarised below.

Since $\hat{H}_\nu^{(2)}(kR)^{1/2}(kR) \equiv (kR)^{1/2}H_{\nu+1/2}^{(2)}(kR)$, eqn. 4.3 becomes

$$\bar{\beta} = \mathrm{j}[1/2kR + H'^{(2)}_{\nu+1/2}(kR)/H^{(2)}_{\nu+1/2}(kR)] \tag{4.5}$$

When $kR \ll |\nu|^2$, we may use the asymptotic expansion of the Hankel function in eqn. 4.5, then an approximate expression for $\bar{\beta}$ is

$$\bar{\beta} \simeq 1 - \nu(\nu+1)/2(kR)^2 \tag{4.6}$$

In the transition region where kR and $|\nu|$ are of the same order, the logarithmic derivative of the Hankel function can accurately be approximated by the Airy function.

$$H'^{(2)}_{\nu+1/2}(kR)/H^{(2)}_{\nu+1/2}(kR) \simeq (2/kR)^{1/3}\, w_1'(t)/w_1(t)$$
$$t = (kR/2)^{2/3}\{(\nu+\tfrac{1}{2})^2/(kR)^2 - 1\}$$

and $w_1(t)$ is the Airy function which equals $B_i(t) - A_i(t)$ and is defined in Reference 178. Computation of the latter is straightforward using small argument and large argument expansions, also given in the reference. Certain simple expressions for $\bar{\beta}$ are then obtainable.

(i) $kR < \mathrm{Re}(\nu+\tfrac{1}{2}) \gg 1$ whence $\mathrm{Re}(-t) \gg 1$

$$\bar{\beta} \simeq (1-z^2)^{1/2} - (\mathrm{j}/2\,kR)z^2/(1-z^2) \tag{4.7}$$

where $z = (\nu+\tfrac{1}{2})/kR$

(ii) $\mathrm{Re}(\nu+\tfrac{1}{2}) > kR \gg 1$ whence $\mathrm{Re}(t) \gg 1$

$$\bar{\beta} \simeq -\mathrm{j}(z^2-1)^{1/2}(1 - z^2/2\,kR(z^2-1)^{3/2}) + [(z^2-1)^{1/2} \exp(-4t^{3/2}/3)] \tag{4.8}$$

Formulae 4.7 and 4.8 reflect clearly the gradual cut off phenomenon which is characteristic of spherical modes.

Modal equation and the hybrid factor

The boundary conditions require that at $\theta = \theta_1$, $E_\phi = 0$ and $-H_\phi/E_R = Y$ where Y is the average slot admittance. A good approximation is given by

$$Y = -\mathrm{j}(\cot kd + 1/2\,kR \sin\theta_1)/\eta_0 \tag{4.9}$$

By using eqns. 4.1 and 4.2, the boundary conditions are

$$p_\nu^m(\theta_1) = -m\bar{\beta}/\bar{\Lambda}\sin\theta_1 \tag{4.10}$$

$$p_\nu^m(\theta_1) = \bar{Y}\nu(\nu+1) = -m\bar{\Lambda}\bar{\beta}/\sin\theta_1 \tag{4.11}$$

where

$$p_\nu^m(\theta_1) \equiv (\partial P_\nu^m(\cos\theta)/\partial\theta)/P_\nu^m(\cos\theta)[\theta = \theta_i] \tag{4.12}$$

and

$$\bar{Y} = -\mathrm{j}Yz_0/kR \tag{4.13}$$

Multiplying eqns. 4.10 and 4.11 together yields

$$p_\nu^m(\theta_1)\,[p_\nu^m(\theta_1) - \bar{Y}\nu(\nu+1)] = m^2\bar{\beta}^2/\sin^2\theta_1 \tag{4.14}$$

Subtracting eqn. 4.11 from eqn. 4.10 yields

$$\bar{\Lambda} - 1/\bar{\Lambda} = \bar{Y}\nu(\nu+1)\sin\theta_1/m\bar{\beta} \tag{4.15}$$

Eqn. 4.14 is the modal equation for ν, while eqn. 4.15 gives the hybrid factor $\bar{\Lambda}$

in terms of the normalised wall admittance $\bar{Y}$. It is noted that when $\bar{Y}=0$, $\bar{\Lambda}=\pm 1$ corresponding to balanced HE or EH modes, respectively. In general, the dependence of the modal eqn. 4.14 on R through $\bar{\beta}$ and possibly $\bar{Y}$ violates the 'separation of variables' assumption. However, near the balanced-hybrid condition $\bar{Y}\sim 0$, numerical solutions of eqn. 4.14 indicate that ν is only very weakly dependent on kR even for $kR \sin\theta_1$ as low as 3. On the other hand, $\bar{\beta}$ can vary widely with kR, taking mainly real values for $kR \gg \nu^2$ and mainly imaginary values for $kR \ll \nu$. For intermediate values of kR, $\bar{\beta}$ is complex. This behaviour, of course, reflects the well known phenomenon describing the gradual cut-off of spherical modes. Of particular interest is the fact that unless the mode is a pure E or H mode ($\bar{\Lambda}=0$ or ∞), the variation of $\bar{\beta}$ with kR results in a varying transverse field distribution as can be seen from eqn. 4.2. As a further consequence, the orthogonality relationship between modes at any given R is not exactly satisfied.

On returning to eqn. 4.14, this has been solved for the first few low order modes under the conditions corresponding to $kR \ll \nu^2$ or $\bar{\beta}\sim 1$ and $\bar{Y}=0$. We designate the solutions in this case by ν_n^0; $n=1, 2\ldots$ Approximate solutions for the HE_{11}, EH_{12} and HE_{12} modes are now discussed.

Approximate solution of the modal equation

For small values of θ_1 and large ν, the Legendre function $P_\nu(\cos\theta_1)$ is well approximated by

$$P_\nu(\cos\theta_1) \simeq J_0((\nu+\tfrac{1}{2})\theta_1) \tag{4.16}$$

where $J_0(\cdot)$ is the Bessel function of zero order. It then follows that

$$P_\nu^1(\cos\theta_1) \equiv \partial P_\nu(\cos\theta_1)/\theta_1 \cong -(\nu+\tfrac{1}{2})J_1((\nu+\tfrac{1}{2})\theta_1) \tag{4.17}$$

and the modal eqn. 4.14, for $m=1$, $\bar{\beta}\sim 1$, becomes

$$\sin\theta_1 p_\nu^1(\theta_1) \cong \theta_1 p_\nu^1(\theta_1) \cong x\,\frac{J_1'(x)}{J_1(x)} = \mp 1 \tag{4.18}$$

where $x=(\nu+\frac{1}{2})\theta_1$.

Solutions of the above equation are simply given by $J_0(x)=0$ or $J_2(x)=0$ for HE_{1n} or EH_{1n} modes, respectively. Using the first zero of $J_2(x)$ and the second zero of $J_0(x)$, we obtain immediately eqns. 4.20 and 4.21.

Using the first zero of $J_0(x)$ we get for the HE_{11} mode $\nu_1^0 \simeq 2{\cdot}405/\theta_1$. This approximation, however, is in relatively large error ($>2{\cdot}5\%$ for $\theta_1>30°$). The formula given by eqn. 4.19 is found to have much less error and is thus considered a better approximation for the HE_{11} mode.

$$\nu_1^0+\tfrac{1}{2} \simeq 2{\cdot}405/(\theta_1\sin\theta_1)^{1/2} \tag{4.19}$$

$$\nu_2^0+\tfrac{1}{2} \simeq 5{\cdot}1356/\theta_1 \tag{4.20}$$

$$\nu_3^0+\tfrac{1}{2} \simeq 5{\cdot}5201/\theta_1 \tag{4.21}$$

They deviate from the exact solutions by no more than 1% for flare angles θ_1 up to 60°. The constants on the RHS of eqns. 4.19 to 4.20 are zeros of the Bessel functions of zero and second orders. The dependence of ν on θ_1 for lowest order modes is to be found in Fig. 4.2(a) while Table 4.1 tabulates ν as a function of θ_1.

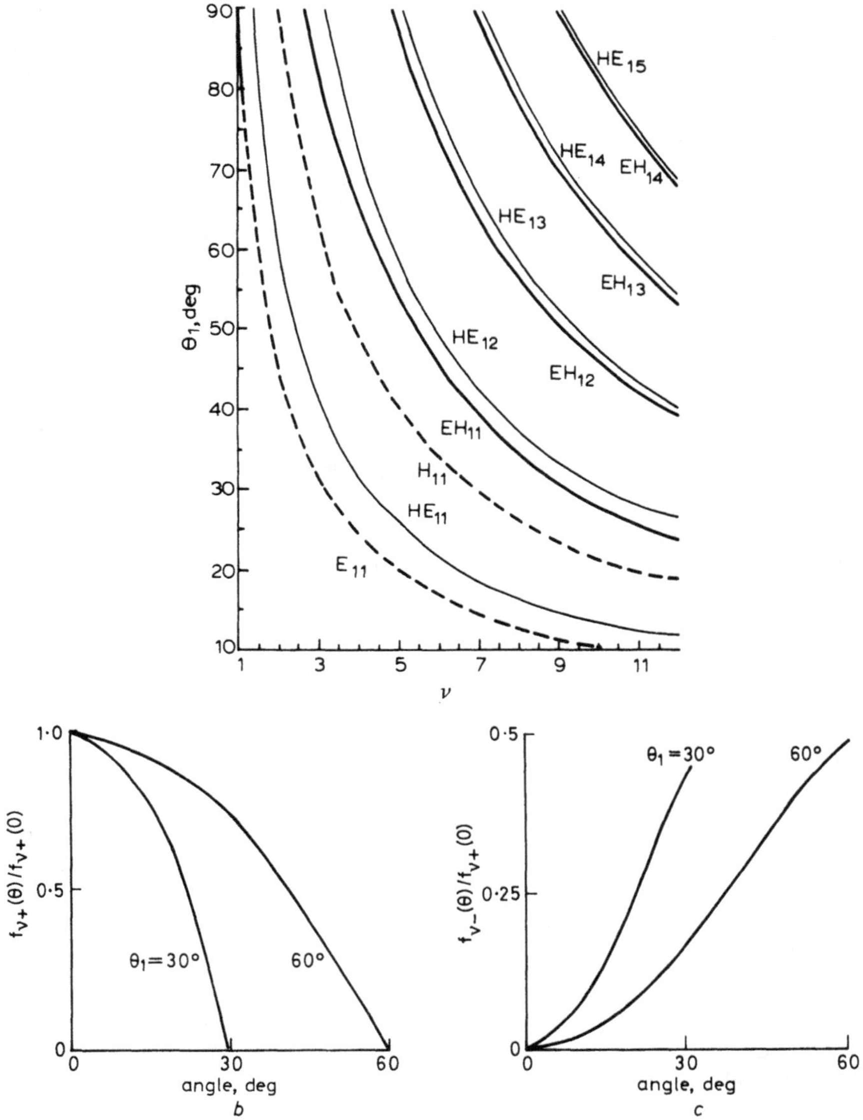

Fig. 4.2 *a Solutions of characteristic equation for HE_{1n} and EH_{1n} modes in corrugated conical horns*

– – – – modes in smooth wall horn

b Function $f_{\nu_+}(\theta)$ as a function of θ

c Function $f_{\nu_-}(\theta)$ as a function of θ

Table 4.1 *First root ν of $f_\nu^1(\theta_1)=0$ for HE_{11} spherical mode*

θ_1^0	ν	θ_1^0	ν	θ_1^0	ν
16	8·1649	41	2·9982	66	1·8126
17	7·6617	42	2·9216	67	1·7851
18	7·2148	43	2·8487	68	1·7585
19	6·8152	44	2·7793	69	1·7327
20	6·4560	45	2·7132	70	1·7078
21	6·1313	46	2·6501	71	1·6837
22	5·8364	47	2·5898	72	1·6604
23	5·5674	48	2·5322	73	1·6379
24	5·3212	49	2·4770	74	1·6161
25	5·0949	50	2·4243	75	1·5949
26	4·8863	51	2·3737	76	1·5745
27	4·6933	52	2·3252	77	1·5546
28	4·5144	53	2·2787	78	1·5354
29	4·3481	54	2·2341	79	1·5168
30	4·1931	55	2·1912	80	1·4988
31	4·0483	56	2·1499	81	1·4813
32	3·9128	57	2·1103	82	1·4643
33	3·7857	58	2·0721	83	1·4479
34	3·6663	59	2·0354	84	1·4319
35	3·5539	60	2·0000	85	1·4164
36	3·4479	61	1·9659	86	1·4014
37	3·3479	62	1·9330	87	1·3869
38	3·2533	63	1·9013	88	1·3728
39	3·1637	64	1·8707	89	1·3591
40	3·0788	65	1·8412	90	1·3458

Finally, it is useful, for the purpose of calculating radiation patterns, to separate the transverse modal fields into their copolar and crosspolar components. We first define appropriate orthogonal unit vectors:

$$\hat{i}_{co} = \cos\phi\,\hat{i}_\theta - \sin\phi\,\hat{i}_\phi \tag{4.22}$$

$$\hat{i}_{cr} = \sin\phi\,\hat{i}_\theta + \cos\phi\,\hat{i}_\phi \tag{4.23}$$

Then, on solving for $\hat{i}_\theta$ and $\hat{i}_\phi$ and substituting in eqn. 4.2 with $m = 1$, the transverse modal fields are recast, apart from a constant, in the form:

$$R\mathbf{E}_{tr} = \tfrac{1}{2}(\bar{\beta}+\bar{\Lambda})f_{\nu+}^1(\theta)\hat{i}_{co} + \tfrac{1}{2}(\bar{\beta}-\bar{\Lambda})f_{\nu-}^1(\theta)\cdot(\cos 2\phi\,\hat{i}_{co} + \sin 2\phi\,\hat{i}_{cr}) \tag{4.24}$$

$$Ry_0\mathbf{H}_{tr} = \tfrac{1}{2}(\bar{\beta}\bar{\Lambda}+1)f_{\nu+}^1(\theta)\hat{i}_{cr} - \tfrac{1}{2}(\bar{\beta}\bar{\Lambda}-1)f_{\nu-}^1(\theta)\cdot(\cos 2\phi\,\hat{i}_{cr} - \sin 2\phi\,\hat{i}_{co}) \tag{4.25}$$

where $f^m_{\nu\pm}(\theta)$ are defined by

$$f^m_{\nu\pm}(\theta) = \partial P^m_\nu(\cos\theta)/\partial\theta \pm mP^m_\nu(\cos\theta)/\sin\theta \quad (4.26)$$

For the HE_{1n} modes, under the balanced hybrid condition $\bar{\Lambda} = 1$, the transverse electric field is mostly in the copolar direction ($\bar{\beta} \sim 1$). The ratio of the crosspolar to copolar component is seen to be maximum at $\phi = \pi/4$ and is given by

$$\frac{E_{cr}}{E_{co}} = [(\bar{\beta} - \bar{\Lambda})/(\bar{\beta} + \bar{\Lambda})]\,[f^1_{\nu-}(\theta)/f^1_{\nu+}(\theta)] \quad (4.27)$$

Graphs of $f^1_{\nu+}(\theta)$ and $f^1_{\nu-}(\theta)$ as a function of θ are shown in Fig. 4.2(b) and (c).

4.3 Radiation from corrugated conical horns by the spherical wave expansion method

The prediction of the radiation fields from conical horns can be accomplished by the Kirchhoff–Huygen, the Laguerre–Gaussian expansion, or the spherical wave expansion methods. We begin by describing the latter method which is as effective in predicting the near fields as the far fields. The spherical wave expansion (SWEX) method was first used with corrugated horns by Clarricoats and Saha [30] who derived expressions for copolar radiation under balanced-hybrid conditions. The method has been generalised by Mahmoud and Clarricoats [95] whose analysis is reproduced below. Simple closed from expressions are given for both copolar and crosspolar radiation fields from conical corrugated horns. The Kirchhoff–Huygens and Laguerre–Gaussian methods are discussed in Sections 4.5 and 4.6.

4.3.1 The spherical wave expansion method

The basic spherical wave functions with satisfy both Maxwell's equations and the radiation condition are given by eqns. 4.28 and 4.29.

$$\bar{m}_{\substack{e\\o},mn} = \{(\mp mP^m_n(\cos\theta)/\sin\theta)^{\sin}_{\cos} m\phi\,\hat{i}_\theta - (\partial P^m_n(\cos\theta)/\partial\theta)^{\cos}_{\sin} m\phi\,\hat{i}_\phi\}\;\hat{h}^{(2)}_n(kR) \quad (4.28)$$

$$\bar{n}_{\substack{e\\o},mn} = (n(n+1)/kR)P^m_n(\cos\theta)^{\cos}_{\sin} m\phi\,\hat{h}^{(2)}_n(kR)\,\hat{i}_R + \{(\partial P^m_n(\cos\theta)/\partial\theta)^{\cos}_{\sin} m\phi\,\hat{i}_\theta \mp (mP^m_n(\cos\theta)/\sin\theta)^{\sin}_{\cos} m\phi\,\hat{i}_\phi\}\;\hat{H}'^{(2)}_n(kR)/kR \quad (4.29)$$

where m and n are integers.

By using these functions, one can expand $\mathbf{E_a}$ and $\mathbf{H_a}$ which are the transverse vector fields at the horn's aperture as

$$\mathbf{E_a} = -\sum_{l=1}^{\infty} (a_l \bar{m}_{o1l} + b_l \bar{n}_{e1l}) \quad (4.30)$$

$$\mathbf{H_a} = (1/jz_0)\sum_{l=1}^{\infty} (a_l \bar{n}_{o1l} + b_l \bar{m}_{o1l} + b_l \bar{m}_{e1l}) \quad (4.31)$$

where a_l and b_l are, as yet, unknown coefficients. In the above, the subscript 1 in the $\bar{m}$ and $\bar{n}$ functions refers to the ϕ variation of the modes which is assumed to be either cos ϕ or sin ϕ, since only the dominant modes are of interest in this analysis. This subscript will be omitted in what follows for convenience.

To obtain the coefficients a_n and b_n, one can use either $\mathbf{E_a}$ alone, $\mathbf{H_a}$ alone, or both $\mathbf{E_a}$ and $\mathbf{H_a}$. Expanding $\mathbf{E_a}$ alone, eqn. 4.30 is cross-multiplied by $\bar{n}_{on}$ and $\bar{m}_{en}$ successively and then integrated over the spherical aperture. Upon using well known orthogonality relationships among the m and n functions one obtains

$$\begin{Bmatrix} a_n \\ b_n \end{Bmatrix} = \pm k^2 \int_0^{\pi}\int_0^{2\pi} \begin{Bmatrix} \bar{n}_{on} \\ \bar{m}_{en} \end{Bmatrix} \times \mathbf{E_a}\, \hat{i}_R R_0^2 \sin\theta\, d\phi d\theta / \gamma_n H_n^{(2)}(kR_0)$$

$$H_n'^{(2)}(kR_0) \tag{4.32}$$

where $\gamma_n = 2\pi n^2(n+1)^2/(2n+1)$. Similarly, for an $\mathbf{H_a}$ expansion, one obtains

$$\begin{Bmatrix} a_n \\ b_n \end{Bmatrix} = \pm\, \mathrm{j} z_0 k^2 \int_0^{\pi}\int_0^{2\pi} \begin{Bmatrix} \bar{m}_{on} \\ \bar{n}_{en} \end{Bmatrix} \times \mathbf{H_a}\, \hat{i}_R R_0^2 \sin\theta\, d\phi d\theta / \gamma_n H_n^{(2)}(kR_0)$$

$$H_n'^{(2)}(kR_0) \tag{4.33}$$

The aperture fields of the HE_1 or EH_1 modes are given by eqns. 4.24–4.25 in the range $0 < \theta < \theta_1$ and they may be taken as zero outside this range. By using these equations along with eqns. 4.22 and 4.23 in eqn. 4.32 we obtain for an $\mathbf{E_a}$ expansion

$$\begin{Bmatrix} a_n \\ b_n \end{Bmatrix} = -(\pi k R_0/2\gamma_n) \begin{Bmatrix} 1/\hat{H}_n^{(2)}(kR_a) \\ 1/\hat{H}_n^{(2)}(kR_a) \end{Bmatrix} [(1+\bar{\Lambda})T_{n+}(\theta_1) \mp (1-\bar{\Lambda})$$

$$T_{n-}(\theta_1)] \tag{4.34}$$

where

$$T_{n\pm}(\theta_1) = \int_0^{\theta_1} f^1_{\nu\pm}(\theta) f^1_{n\pm}(\theta) \sin\theta\, d\theta \tag{4.35}$$

This integral can be expressed in closed form:

$$\int_0^{\theta_1} f^m_{\nu\pm}(\theta) f^m_{w\pm}(\theta) \sin\theta\, d\theta = \{\sin\theta_1/(\nu(\nu+1) - w(w+1))\}$$

$$\{\nu(\nu+1) P^m_\nu(\cos\theta_1) f^m_{w\pm}(\theta_1) - w(w+1) P^m_w(\cos\theta_1) f^m_{\nu\pm}(\theta_1)\} \tag{4.36}$$

where $f^m_{\nu\pm}(\theta)$ is defined in eqn. 4.26.

When w approaches ν, the above integration is obtained by taking the limit $w \to \nu$ of the RHS.

$$\int_0^{\theta_1} [f_{\nu\pm}^m(\theta)]^2 \sin\theta d\theta = P_\nu^m(\cos\theta_1) f_\nu^m(\theta_1)\sin\theta_1 - [P_\nu^m(\cos\theta_1)$$

$$\partial f_{\nu\pm}^m(\theta_1)/\partial\nu - f_{\nu\pm}^m(\theta_1)\,\partial P_\nu^m(\cos\theta_1)/\partial\nu]\ \frac{\nu(\nu+1)\sin\theta_1}{2\nu+1} \tag{4.37}$$

Similarly, for an $\mathbf{H_a}$ expansion one obtains

$$\begin{Bmatrix} a_n \\ b_n \end{Bmatrix} = -(\pi kR/2\gamma_n)\begin{Bmatrix} -\mathrm{j}/\hat{H}_n'^{(2)}(kR_0) \\ \mathrm{j}/\hat{H}_n^{(2)}(kR_0) \end{Bmatrix} [(1+\bar{\Lambda})T_{n+}(\theta_1) \mp (1-\bar{\Lambda})$$

$$T_{n-}(\theta_1)] \tag{4.38}$$

In eqns. 4.34 and 4.38 the approximation $\bar{\beta} \simeq 1$ has been used, which is valid for large apertures where $kR_0 \gg \nu^2$.

At this point, one notices that these equations are not identical. This is because $\mathbf{E_a}$ and $\mathbf{H_a}$ are not quite exact since they are assumed to be zero for $\theta > \theta_1$; also the horn air discontinuity is neglected. However, for large apertures ($kR_0 \gg \nu^2$) one expects that the lower order terms in the expansion will be the most significant in an expansion of the aperture field. For these terms $kR_0 \gg n^2$ and $\hat{H}_n'^{(2)}(kR_0) \cong -\mathrm{j}\,\hat{H}_n^{(2)}(kR_0)$, then under this approximate condition, the above equations are indeed identical. Now using this condition and substituting from either eqn. 4.34 or eqn. 4.38 in eqn. 4.30, the electric field $\mathbf{E_p}$ in the radiation zone is expressed in terms of its copolar and crosspolar components:

$$\mathbf{E_p}(R,\theta,\phi) = \{\tfrac{1}{2}(1+\bar{\Lambda})\hat{i}_{co}S_+(\theta,\theta_1) + \tfrac{1}{2}(1-\bar{\Lambda})(\hat{i}_{co}\cos 2\phi + \hat{i}_{cr}\sin 2\phi)$$

$$S_-(\theta,\theta_1)\exp(-\mathrm{j}kR)/(R/R_0) \tag{4.39}$$

$$S_\pm(\theta,\theta_1) = \sum_{n=1}^{\infty} C_{n\pm}(\theta_1) f_{n\pm}^1(\theta)\mathrm{j}^{n+1}/\hat{h}_n^{(2)}(kR_0) \tag{4.40}$$

$$C_\pm(\theta_1) = T_{n\pm}(\theta_1)\cdot\pi/\gamma_n$$

$$= \frac{2n+1}{2n^2(n+1)^2}\cdot\frac{\sin\theta_1}{n(n+1)-\nu(\nu+1)}\cdot[n(n+1)P_n^1(\cos\theta_1)$$

$$f_{\nu\pm}^1(\theta_1) - \nu(\nu+1)P_\nu^1(\cos\theta_1)\,f_{n\pm}^1(\theta_1)] \tag{4.41}$$

A slightly different formula from eqn. 4.39 is obtained by averaging out the coefficients of eqns. 4.34 and 4.38. This results in a correction term s_n inside the summation of eqn. 4.40 where

$$s_n = [1 + \hat{H}_n^{(2)}(kR_0)/\mathrm{j}\hat{H}_n'^{(2)}(kR_0)]/2 \tag{4.42}$$

Again it is noted that for the lower order terms for which $kR_0 \gg n^2$, s_n tends to unity. Numerical results confirm that for sufficiently large apertures, e.g. $kR_0 \sin\theta_1 \gg 15$, the correction term in eqn. 4.42 does not change the radiation pattern by more than a fraction of 1 dB.

4.3.2 Frequency dependence of crosspolar radiation field

From eqn. 4.39 the crosspolar radiation field attains its maximum in the $\phi = 45°$ plane. Relative to the boresight copolar radiation level, the crosspolar field in this plane is given by

$$C(\theta,\theta_1) = [(1-\bar{\Lambda})/(1+\bar{\Lambda})]\, S_-(\theta,\theta_1)/S_+(0,\theta_1) \qquad (4.43)$$

The key factor that determines the frequency dependence of C is that which involves $\bar{\Lambda}$. Looking back at eqn. 4.15 and noting that for all practical cases $\bar{Y}\nu(\nu+1)\sin\theta_1 \ll 1$, $\bar{\Lambda}$ is approximated by $1 + \frac{1}{2}\bar{Y}\nu(\nu+1)\sin\theta_1$ for HE_{1n} modes. Using eqns. 4.5 and 4.13 for $\bar{Y}$, eqn. 4.43 then takes the form

$$C(\theta,\theta_1) \simeq \frac{-\nu(\nu+1)\sin^2\theta_1}{4kr_{1a}} \, \frac{S_-(\theta,\theta_1)}{S_+(0,\theta_1)} \, [-\cot ks + 1/2kr_{1a}] \qquad (4.44)$$

which shows that the frequency variation of C is mainly controlled by the resonant behaviour of the slots. So let f_0 be the frequency at which $k_0 s = \pi/2$ or $s = \lambda_0/4$ and define $\delta = (f/f_0 - 1)$, then eqn. 4.44 becomes for the HE_{11} mode:

$$C(\theta,\theta_1) = [(2{\cdot}405)^2 \sin\theta_1/\theta_1 - \sin^2\theta_1/4] \, \frac{S_-(\theta,\theta_1)}{S_+(0,\theta_1)}$$

$$\left[\frac{\tan \pi\delta/2}{1+\delta} - \frac{1}{2k_0 r_{1a}(1+\delta)^2}\right] \Big/ 4k_0 r_{1a} \qquad (4.45)$$

where eqn. 4.19 has been used as an approximation for ν. In this form the cross-polar ratio is expressed as flare-angle dependent terms multiplied by a frequency-dependent term. It is noted that C decreases steadily with increasing horn aperture. The latter actually determines the bandwidth of the horn as is demonstrated in Fig. 4.3 where C is shown as a function of δ with $k_0 r_{1a}$ as a parameter. The dependence of C on the flare angle θ_1 is almost as $\sin\theta_1/\theta_1$ as is seen from eqn. 4.45 and hence it is a slowly varying function of the flare angle in the range of interest. The radiation patterns of copolar and crosspolar fields are given by $S_\pm(\theta,\theta_1)$, respectively. These are plotted in Fig. 4.4 for 30° and 45° horns having the same aperture: $kr_{1a} = 15$. The levels in the figure correspond to a frequency deviation of + 10% from f_0 (i.e. $\delta = 0.1$) and for both horns normalization is made relative to the on-axis copolar radiation level $S_+(0, \theta_1)$. The crosspolar patterns of the EH_{12} mode in the same horns of Fig. 4.3 are shown in Fig. 4.4.

The gain G of the horn supporting a given mode can be deduced from eqn. 4.39. For simplicity let us consider the case $\bar{\Lambda} = \pm 1$ corresponding to balanced HE or EH modes.

$$G = 4\pi R^2\, |\hat{E}_p|^2 / y_0 P \qquad (4.46)$$

$$P = 2\pi R_0^2 \int_0^{\theta_1} [f_{\nu\pm}^1(\theta)]^2 \sin\theta d\theta / y_0$$

Upon using eqn. 4.39, this is reduced to

$$G = 2|\hat{S}_\pm|^2 \Big/ \int_0^{\theta_1} [f_{\nu\pm}^1(\theta)]^2 \sin\theta d\theta \tag{4.47}$$

A further reduction is obtained by using eqn. 4.37 and the modal equation $f_{\nu\pm}^1(\theta_1) = 0$ under the balanced mode condition.

$$G = \frac{-2(2\nu+1)}{\nu(\nu+1)\sin\theta_1} \frac{|S_\pm|^2}{P_\nu^1(\cos\theta_1)\dfrac{\partial}{\partial\nu} f_{\nu\pm}^1(\theta_1)} \tag{4.48}$$

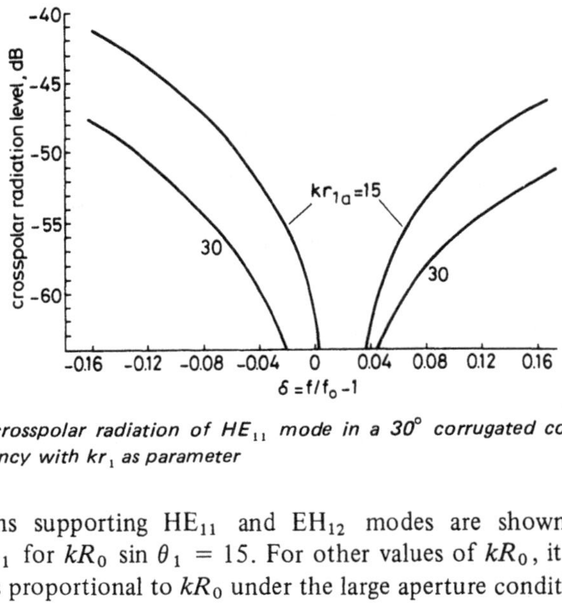

Fig. 4.3 *Peak crosspolar radiation of HE_{11} mode in a 30° corrugated conical horn against frequency with kr_1 as parameter*

Gains of horns supporting HE_{11} and EH_{12} modes are shown in Fig. 4.5 as functions of θ_1 for $kR_0 \sin\theta_1 = 15$. For other values of kR_0, it is useful to note that the gain is proportional to kR_0 under the large aperture condition.

4.4 Generation of higher order modes

Excitation of the dominant HE_{11} mode in a corrugated conical horn inevitably generates other higher order modes. Of these the EH_{12} mode most seriously affects the purity of the polarisation of the radiated fields. This is so since near the balanced-hybrid frequency the HE_{1n} modes for $n > 1$ make negligible contribution to crosspolar radiation. Generation of the EH_{12} mode can occur at the throat, also along the horn due to conversion from the HE_{11} mode. Mode conversion along the horn occurs either when the normalised slot admittance $\bar{Y}$ is varying or if the horn has a non-linear profile or both. Dragone [56] has derived a simple formula for mode conversion in a narrow flare-angle horn under the large aperture

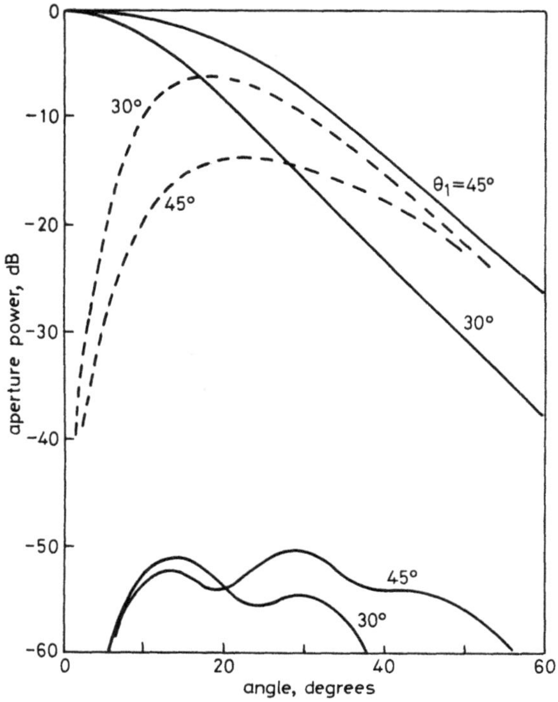

Fig. 4.4 *Copolar and crosspolar radiation patterns of HE_{11} mode at a frequency of $f = 1{\cdot}1 f_0$, $kr_1 = 15$, $\theta_1 = 30, 45$*

---- Copolar patterns for EH_{12} mode

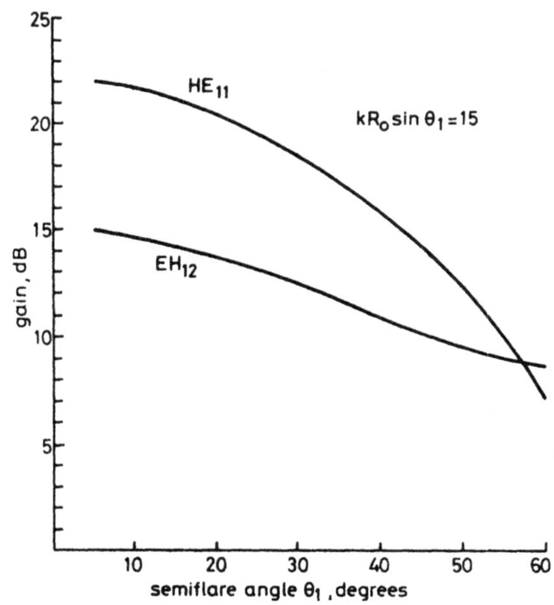

Fig. 4.5 *Gain of HE_{11} and EH_{12} modes against flare angle θ_1*

condition. In this section his restrictions are removed and the generation of the EH_{12} mode is deduced in a horn of arbitrary flare angle both at the throat discontinuity and due to HE_{11} mode conversion along the horn. It is shown that while Dragone's cylindrical mode analysis overestimates the crosspolar contribution due to mode conversion along the horn, the throat contribution is significant, especially for wide flare-angle horns.

4.4.1 Scattering at an abrupt discontinuity

Consider the canonical problem of scattering at a junction between two corrugated conical horns characterised by semi-flare angles and normalised wall admittances $(\psi_1, \bar{Y}_a)$ and $(\theta_1, \bar{Y}_b)$. The geometry and definition of symbols are illustrated in Fig. 4.6. The dominant mode is incident from the LHS and reflected and transmitted modes are excited at the junction. Let $(\mathbf{e}_n^a, \mathbf{h}_n^a)$ be the transverse vector field of the forward nth mode on the LHS of the junction and $(\mathbf{e}_n^b, \mathbf{h}_n^b)$ the corresponding fields on the RHS. A reflected mode on the LHS has the fileds $(\mathbf{e}_n^{a*}, -\mathbf{h}_n^{a*})$ as can be deduced from eqn. 4.2 by noticing that for a reflected mode the following changes apply: $\bar{\beta} \rightarrow \bar{\beta}^*$ and $\bar{\Lambda} \rightarrow \bar{\Lambda}^*$, $\nu \rightarrow \nu^*$ where the symbol * signifies the complex conjugate operation.

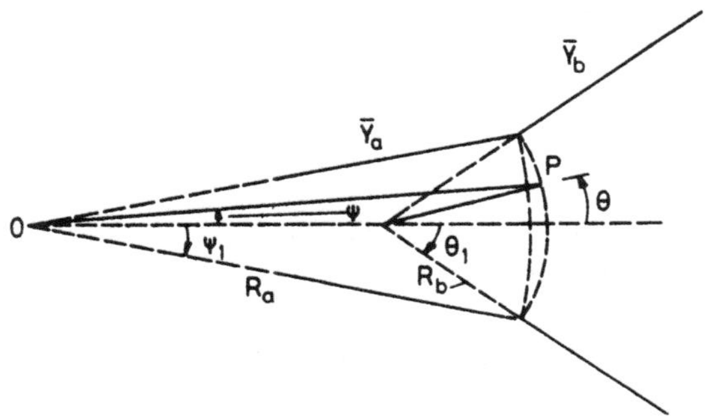

Fig. 4.6 *Junction between two spherical cones with flare angles ψ_1 and θ_1, radii R_a and R_b and wall admittances $\bar{Y}_a$ and $\bar{Y}_b$*

Now matching the transverse fields on both sides of the junction at $R = R_b$ and $0 < \theta < \theta_1$:

$$\mathbf{e}_1^a \exp(-j\beta_a \Delta(\theta) + \sum_{l=1}^{\infty} \Gamma_l \mathbf{e}_l^{a*} \exp(j\beta_a \Delta(\theta)) = \sum_{l=1}^{\infty} T_l \mathbf{e}_l^b \tag{4.49}$$

$$\mathbf{h}_1^a \exp(-j\beta_a \Delta(\theta)) - \sum_{l=1}^{\infty} \Gamma_l \mathbf{h}_l^{a*} \exp(j\beta_a \Delta(\theta)) = \sum_{l=1}^{\infty} T_l \mathbf{h}_l^b \tag{4.50}$$

where $\beta_a = \bar{\beta}_a k$ and $\Delta(\theta)$ is the distance between the two spherical caps $R = R_a$

and $R = R_b$ (see Fig. 4.6) and is given from the geometry by:

$$\Delta(\theta) = R_a[1 + 2(\sin\psi_1/\sin\theta_1)(\cos\psi_1 - \sin\psi_1 \cot\theta_1)(\cos\theta - \cos\theta_1)]^{1/2} - R_a \tag{4.51}$$

The special case $\psi_1 = 0$, corresponding to a cylindrical guide, is of interest. In this case $\Delta(\theta)$ reduces to

$$\Delta(\theta)|_{\psi_1=0} = R_b(\cos\theta - \cos\theta_1) \tag{4.52}$$

To obtain the reflection and transmission coefficients eqns. 4.49 and 4.50 are vector cross-multiplied by $\mathbf{h}_n^b$ and $\mathbf{e}_n^b$, respectively, and then integrated over the spherical cap $R = R_b, 0 < \theta < \theta_1$ to give

$$\langle \mathbf{e}^a \exp(-j\beta_a\Delta(\theta)), \mathbf{h}_n^b\rangle + \sum_l \Gamma_l\langle \mathbf{e}^a \exp(j\beta_a\Delta(\theta)), \mathbf{h}_n^b\rangle$$

$$= T_n + \sum_{l\neq n} T_l\langle \mathbf{e}_l^b, \mathbf{h}_n^b\rangle \tag{4.53}$$

$$\langle \mathbf{e}_n^b, \mathbf{h}^a \exp(-j\beta_a\Delta(\theta))\rangle + \sum_l \Gamma_l\langle \mathbf{e}_n^b, \mathbf{h}_l \exp(j\beta_a\Delta(\theta))\rangle =$$

$$= T_n + \sum_{l\neq n} T_l\langle \mathbf{e}_n^b, \mathbf{h}_l^b\rangle \tag{4.54}$$

for $n = 1, 2 \ldots$ where the normalisation $\langle \mathbf{e}_n^b, \mathbf{h}_n^b\rangle = 1$ has been assumed. $\langle \mathbf{e}^a, \mathbf{h}^b\rangle$ represents the inner product, given by

$$\langle \mathbf{e}^a, \mathbf{h}^b\rangle = \int_0^{2\pi} d\phi \int_0^{\theta_1} \sin\theta \, d\theta \, (R\mathbf{e}^a \times R\mathbf{h}^b)\cdot \hat{i}_R$$

Now following Dragone [56] a simple solution of the above infinite set of equations is obtained by assuming that the coefficients $T_n, n > 2$ and $R_n, n > 1$ are all small quantities relative to T_1. This is usually the case for a weak discontinuity as in all cases considered here. Since the modes on either side of the discontinuity are not exactly orthogonal in our case, the products $\langle \mathbf{e}_n^b, \mathbf{h}_n^b\rangle$, $l \neq n$ are non-zero, but they are small relative to unity. Taking this into account and neglecting quantities of second order of smallness, one obtains:

$$T_1 \simeq \tfrac{1}{2}[\langle \mathbf{e}_1^a \exp(-j\beta_a\Delta), \mathbf{h}_1^b\rangle + \langle \mathbf{e}_1^b, \mathbf{h}_1^a, \exp(-j\beta_a\Delta)\rangle] \tag{4.55}$$

$$\Gamma_1 \simeq -\frac{\langle \mathbf{e}_1^a \exp(-j\beta_a\Delta), \mathbf{h}_1^b\rangle - \langle \mathbf{e}_1^b, \mathbf{h}_1^a \exp(-j\beta_a\Delta)\rangle}{\langle \mathbf{e}_1^{*a} \exp(j\beta_a^*\Delta), \mathbf{h}_1^b\rangle + \langle \mathbf{e}_1^b, \mathbf{h}_1^{*a} \exp(j\beta_a^*\Delta)\rangle} \tag{4.56}$$

$$T_n = \tfrac{1}{2}[\langle \mathbf{e}_1^a \exp(-j\beta_a\Delta), \mathbf{h}_n^b\rangle + \langle \mathbf{e}_n^b, \mathbf{h}_1^a \exp(-j\beta_a\Delta)\rangle] - \tfrac{1}{2}T_1[\langle \mathbf{e}_1^b, \mathbf{h}_n^b\rangle + \langle \mathbf{e}_n^b, \mathbf{h}_1^b\rangle]; n > 1 \tag{4.57}$$

It is constructive to note that these results reduce to those of Dragone's [56] when $\langle \mathbf{e}_n^b, \mathbf{h}_l^b \rangle = 0$, $n \neq l$ and $\Delta = 0$.

4.4.2 Higher order mode generation at the throat

The throat junction is composed of a cylindrical waveguide with smooth walls on one side and the conical corrugated horn on the other. This corresponds to the case treated in the previous section with $\psi_1 = 0$ and $Y_a = \infty$. The cylindrical waveguide normally carries the dominant TE_{11} mode which is incident on the junction. The HE_{11} is excited in the horn along with other higher order modes. A numerical example is shown in Fig. 4.7 for a 30° horn with a slot depth of $\lambda_0/4$ at the centre frequency for which $k_0 r_{1t} = 3$. The transmission coefficients T_1 for the HE_{11} mode and T_2 for the EH_{12} mode and the reflection coefficient Γ_1 of the TE_{11} mode are plotted in decibels as a function of the frequency parameter δ. It is interesting to note that the peak of $|T_2|$ occurs when $|\beta|$ of the EH_{12} mode is about a minimum, i.e. when this mode is in its transition region between a purely propagating state and purely cut-off state.

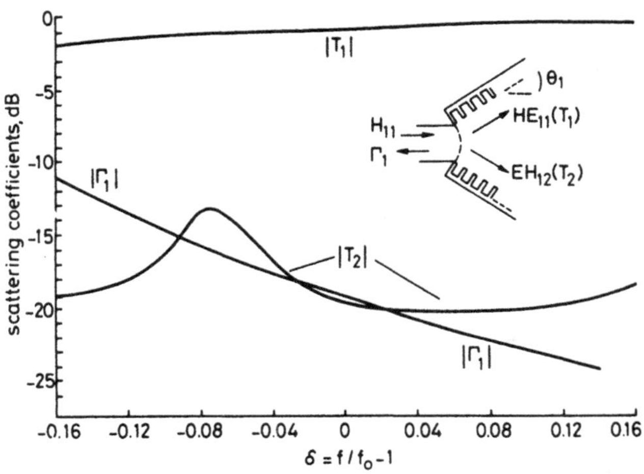

Fig. 4.7 *Reflection and transmission coefficients as a function of frequency for junction between smooth wall waveguide and conical horn, $kr_1 = 3{\cdot}3$, $\theta_1 = 30°$*

As the EH_{12} mode, which is generated at the throat, travels to the mouth of the horn, it suffers a loss which depends on the length of the horn over which the mode is effectively in a cut-off state. This loss is thus a function of both the frequency and the flare angle and is plotted in Fig. 4.8 versus δ for three corrugated horns with $\theta_1 = 12°$, 30° and 45° but with each having the same aperture at the throat and mouth. The net crosspolar level of the EH_{12} mode is then obtained and is shown plotted for $\theta_1 = 30°$ in Fig. 4.9.

4.4.3 Mode conversion along a horn with constant flare angle

The second mechanism responsible for the generation of the EH_{12} mode is the continuous conversion of the HE_{11} mode as this mode travels along the horn. As previously stated it is caused by either a change of $\bar{Y}$ or θ_1 or both. While these two causes can be treated analytically starting from the results of the previous

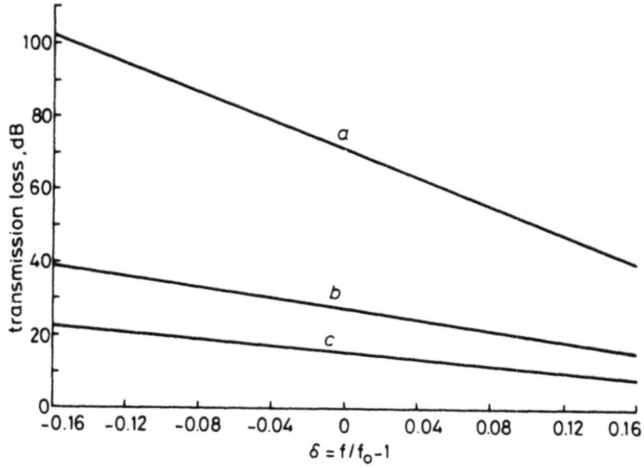

Fig. 4.8 *Transmission loss of EH_{12} mode from the throat to the aperture of a horn as a function of frequency, $kr_{1t} = 3$ and $kr_{1a} = 15$*
a $\theta_1 = 12°$
b $\theta_1 = 30°$
c $\theta_1 = 45°$

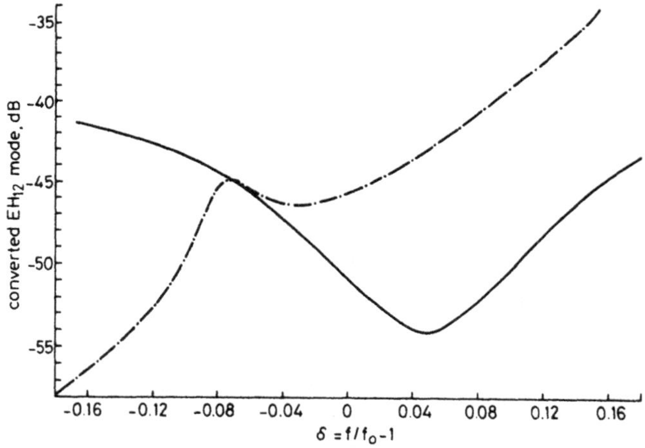

Fig. 4.9 *Levels of EH_{12} mode at the aperture due to the mode generated at the throat and the mode converted along the horn*
- - · - - mode generated at throat
——— mode converted along horn. Parameters as Fig. 4.8

subsection, we shall concentrate here on mode conversion along a straight horn, i.e. a horn with constant flare angle θ_1, so that the only cause of mode conversion is the variation of $\bar{Y}$ with R.

Since the expected levels of converted modes are quite low, certain simplifying assumptions can be made. All reflected modes are neglected and the reconversion of the higher order modes is not considered. In the following it is convenient to adopt the normalisation $\langle \mathbf{e_n}, \mathbf{h_n} \rangle = 1$. This implies that the modal fields in eqn. 4.4 be multiplied by $1/N_n$ where $N_n = \langle \mathbf{E}_n, \mathbf{H}_n \rangle$, and

$$N_n^2 = \langle \mathbf{E}_n, \mathbf{H}_n \rangle / y_0 = (\pi/2)[(\bar{\beta}_n + \bar{\Lambda}_n)(\bar{\beta}_n \bar{\Lambda}_n + 1)L_n^+ - (\bar{\beta}_n - \bar{\Lambda}_n)$$
$$(\bar{\beta}_n \bar{\Lambda}_n - 1)L_n^-] \tag{4.58}$$

where

$$L_n^{\pm} = \int_0^{\theta_1} [f_{n\pm}^1(\theta)]^2 \sin\theta \, d\theta \tag{4.59}$$

and is obtainable from eqn. 4.37.

Now let the complex amplitudes of the various modes be given by $A_n(R)$, $n = 1, 2 \ldots$ At the throat $R = R_0$, $A_1 = 1$ and $A_n = 0$, $n > 1$. Considering mode scattering in an infinitesimal region between $R = R$ and $R = R + dR$ where $\bar{Y}$ changes by $d\bar{Y} = (d\bar{Y}/dR)dR$, the transmission coefficients are obtained by using the results of the previous subsection. A differential relation $A_n(R)$, $n > 1$ is

$$dA_n(R)/dR = -jk\bar{\beta}_n(R)A_n(R) + \frac{\partial T_n}{\partial R_1} A_1(R); n > 1 \tag{4.60}$$

where dT_n is obtainable from eqn. 4.57. After some algebra, an explicit expression for dT_n is

$$dT_n(R) = \frac{\pi p_{\nu 1} p_{\nu n}}{N_1 N_n (\nu_n - \nu_1)} \left[\sin\theta_1 \, \nu_1 \nu_n \left(\frac{-d\bar{Y}}{\bar{\beta}_1} + \frac{\bar{Y}}{\bar{\beta}_1} \frac{d\bar{\beta}_1}{\bar{\beta}_1} \right) \right.$$
$$\left. + \frac{1 + \bar{\Lambda}_1 \bar{\Lambda}_n}{\bar{\Lambda}_1 \bar{\Lambda}_n} (\Lambda_n \nu_n d\bar{\beta}_1 + \nu_1 (1 - \bar{\beta}_n) d\bar{\Lambda}_1) \right]$$
$$+ \frac{\pi p_{\nu 1} p_{\nu n}}{2 N_1 N_n} (1 - \bar{\beta}_n)[(1 - \bar{\beta}_1) d\bar{\Lambda}_1 + (\bar{\Lambda}_1 + \bar{\Lambda}_n) d\bar{\beta}_1] \tag{4.61}$$
$$- \frac{Q_{1n}}{N_1 N_n} \frac{dN_1}{N_1}$$

$$Q_{ab} = \frac{\pi p_{\nu a} p_{\nu b}}{(\nu_b - \nu_a)} [\beta_{ab} \sin\theta_1 \, \nu_a \nu_b (\bar{Y}_a / \bar{\beta}_a - \bar{Y}_b / \bar{\beta}_b) + \Delta_{ab}] \tag{4.62}$$

where

$$\Delta_{ab} = (\beta_{ab} / \bar{\Lambda}_a \bar{\Lambda}_b)[(\zeta_a \bar{\Lambda}_b \nu_b - \zeta_b \bar{\Lambda}_a \nu_a)(1 + \bar{\Lambda}_a \bar{\Lambda}_b)] \, ¼$$
$$+ \zeta_a \zeta_b \bar{\Lambda}_{ab} (\nu_b - \nu_a) \tag{4.63}$$

and $\zeta_{a,b} = 1 - \bar{\beta}_{a,b}$, where all quantities are to be calculated at R, while the quantity

$d\bar{Y}$ is obtained from $d\bar{Y} = \bar{Y}(R + dR) - \bar{Y}(R)$ and likewise for all other differentials.

The amplitude of the dominant mode $A_1(R)$ is generally given by

$$A_1(R) = \exp\left\{-jk\int_{R_0}^{R} \bar{\beta}_1'(R)dR\right\} \quad (4.64)$$

where $\bar{\beta}_1'$ should, in principle, be different from $\bar{\beta}_1$ in order to account for the loading of the higher order modes. However, in our case, because of weak coupling, it is reasonable to take $\bar{\beta}_1' = \bar{\beta}_1$. Now integrating eqn. 4.60, one gets

$$A_n(R) = \int_{R_0}^{R} A_1(R')(dT_n/dR')\exp\left\{-jk\int_{R'}^{R} \bar{\beta}_n(R)dR\right\} dR' \quad (4.65)$$

Substituting from eqn. 4.64

$$A_n(R) = \exp(-j\phi_n)\int_{R_0}^{R} (dT_n/dR')\exp\left\{-jk\int_{R_0}^{R'} (\bar{\beta}_1 - \bar{\beta}_n)dR\right\} dR' \quad (4.66)$$

where

$$\phi_n = k\int_{R_0}^{R} \bar{\beta}_n(R)\,dR \quad (4.67)$$

Now eqn. 4.66 along with eqn. 4.61 give the amplitude of a converted mode at any point along the horn. The integration can be performed numerically. However, a better way is to approximate the quantity $\{(kR)^2(\bar{\beta}_1 - \bar{\beta}_n)\}$ as a linear function in $(1/kR)$ and (dT_n/dkR) as a polynomial in the same variable whereby the integration can be readily expressed as sum of complementary error functions. In certain cases it is necessary to break the interval of integration into two or more subintervals where the above approximations are applied in each subinterval. This procedure ensures an accurate computation of the integral.

Some insight into the mode conversion process to the EH_{12} mode can be gained by examining eqn. 4.61 under the conditions $kR \gg \nu_1, \nu_2$ and $\bar{Y} \ll 1$. Then the term involving $d\bar{Y}$ is dominant and dT_2 becomes

$$\frac{-\pi p_{\nu 1} p_{\nu 2}}{N_1 N_2(\nu_2 - \nu_1)} \sin\theta_1 \nu_1 \nu_2\, d\bar{Y} \quad (4.68)$$

$$N_n^2 \simeq \pi[P_\nu^1(\cos\theta_1)]^2 \nu(\nu + 1)\sin\theta_1 \quad (4.69)$$

Using eqn. 4.69 for N_1 and N_2 under balanced hybrid conditions, and eqn. 4.19 and 4.20 for ν_1 and ν_2, dT_2 further simplifies to the following with $x_0 \simeq 2{\cdot}405$ and $x_2 \simeq 5{\cdot}1356$:

$$\frac{x_0 x_2(\theta_1 \sin\theta_1)^{-1/2}}{x_2^2 - x_0^2\theta_1/\sin\theta_1}\, d\bar{Y} \quad (4.70)$$

Now $d\bar{Y}$ is nearly proportional to $(1/kR)$ which in turn is proportional to $\sin\theta_1$ assuming the horn's plane aperture is kept constant. The net result is that dT_2 is very weakly dependent on the flare angle θ_1. Now, turning attention to eqn. 4.66, and identifying the term in brackets as ϕ_{12}, the real part is the difference of phase delays of the first and second modes along the horn. This is obviously minimum for the shortest horn ($\theta_1 = 90°$), provided that the throat and mouth apertures are kept constant. If this phase difference is less than π for $\theta_1 = 90°$, the magnitude of $A_2(R_m)$ will tend to increase if θ_1 is reduced, it attains a maximum when this phase is $\sim\pi$ and goes into a damped oscillation with a further decrease of θ_1. This behaviour is illustrated in Fig. 4.10, which shows the level of EH_{12} mode converted along a conical horn versus the semi-flare angle θ_1 with $f/f_0 = 1{\cdot}1$. Also shown in the figure, curve b, is the upper bound predicted by using the cylindrical mode theory of Dragone from which it can be seen that this theory overestimates the role of mode conversion in wide flare-angle horns.

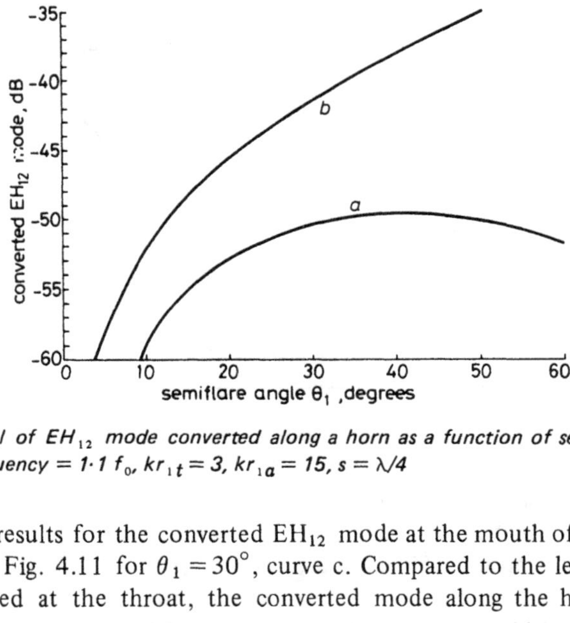

Fig. 4.10 *Level of EH_{12} mode converted along a horn as a function of semi-flare angle θ_1, frequency = $1{\cdot}1\,f_0$, $kr_{1t} = 3$, $kr_{1a} = 15$, $s = \lambda/4$*

Numerical results for the converted EH_{12} mode at the mouth of a horn versus δ are shown in Fig. 4.11 for $\theta_1 = 30°$, curve c. Compared to the level of the EH_{12} mode generated at the throat, the converted mode along the horn, curve b, is lower over a broad band of frequencies but the reverse would be true for narrow flare angles. For the 30° horn the mode generated at the throat dominates at the higher end of the frequency band while the converted mode dominates at the lower end. This behaviour is mainly attributed to the higher transmission loss of the throat generated mode in the narrower horn.

One effective way of reducing the level of the EH_{12} mode in wider flare-angle horns is to start the horn at a smaller throat aperture. This increases the total transmission loss for an EH_{12} mode which is generated at the throat. But, eventually there would be an increase in the level of the EH_{12} mode generated at the throat. A remedy for reducing the return loss in narrow flare horns is to use the $\lambda/2$ deep corrugations at the throat. The corrugation depth is gradually reduced to $\lambda/4$ at the mouth of the horn (see Section 5.5). This procedure can also result in some reduction of the EH_{12} mode.

The overall crosspolar peak radiation level of a 30° horn with corrugations of $\lambda/4$ depth at mid-band is shown in Fig. 4.11. For this corrugated horn $k_0 r_{1t} = 3$ and $k_0 r_{1a} = 15$. It is seen that, in this case, mode conversion along the horn is insignificant over all the band. It is dominated by the mode generated at the throat at the higher frequencies and by the intrinsic crosspolarisation of the HE_{11} mode curve a, at the lower frequencies. Experimental data for the same horn are shown for comparison and a satisfactory agreement with the overall theoretically predicted level is observed.

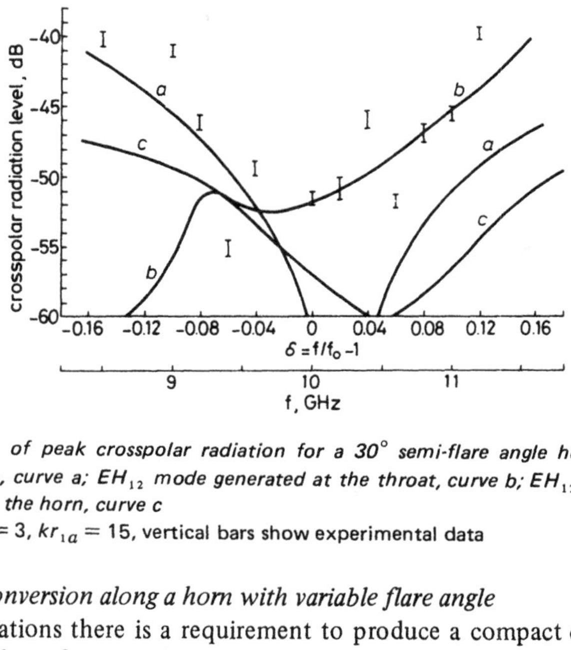

Fig. 4.11 *Level of peak crosspolar radiation for a 30° semi-flare angle horn due to HE_{11} mode, curve a; EH_{12} mode generated at the throat, curve b; EH_{12} mode converted along the horn, curve c*
$kr_{1t} = 3$, $kr_{1a} = 15$, vertical bars show experimental data

4.4.4. Mode conversion along a horn with variable flare angle

In some applications there is a requirement to produce a compact corrugated feed with a plane phase front in the aperture. This must be of sufficient size in terms of wavelengths, so as to produce a narrow beamwidth pattern. One way to create such a feed is to utilise the profiled horn of Fig. 4.12(a). Such horns have been fabricated but have been found to contain higher-order modes, generated because of the changing profile. Generally their level exceeds that which arises in a linear profile horn as discussed in the previous section.

Based on the work of Mahmoud [177], we shall now determine the effect of mode conversion in a profiled horn as shown in Fig. 4.12(a). It is assumed that the dominant HE_{11} mode exists in the throat at $z = 0$, also that the walls can be modelled by a constant surface admittance. In practice the admittance may be varied as well as the flare angle but for purpose of discussion the effects are separated here.

From eqn. 4.66 we have

$$A_n(L)/A_1(L) = \exp(-\mathrm{j}(\phi_n - \phi_1)) \int_{z=0}^{L} \mathrm{d}C_n(z) \exp \left[-\mathrm{j} \int_0^z \{\beta_1(z') - \beta_n(z')\}(\mathrm{d}s/\mathrm{d}z')\mathrm{d}z' \right] \quad (4.71)$$

where

$$\phi_i = \int_0^L \beta_i(z')\,\mathrm{d}s, \; i = 1 \text{ or } n$$

$$\beta_i = \text{the modal complex phase constant}$$

$$\mathrm{d}s = \mathrm{d}z'(1 + (\mathrm{d}r/\mathrm{d}z')^2)^{1/2}$$

and $\mathrm{d}C_n(z)$ is the differential conversion from the dominant mode to the nth mode per unit amplitude of the dominant mode in the region bounded by $z - \mathrm{d}z$ and z. Now, the crux of the problem is to derive $\mathrm{d}C_n(z)$ in a closed form in terms of the local profile parameters and the local modal parameters of the modes involved.

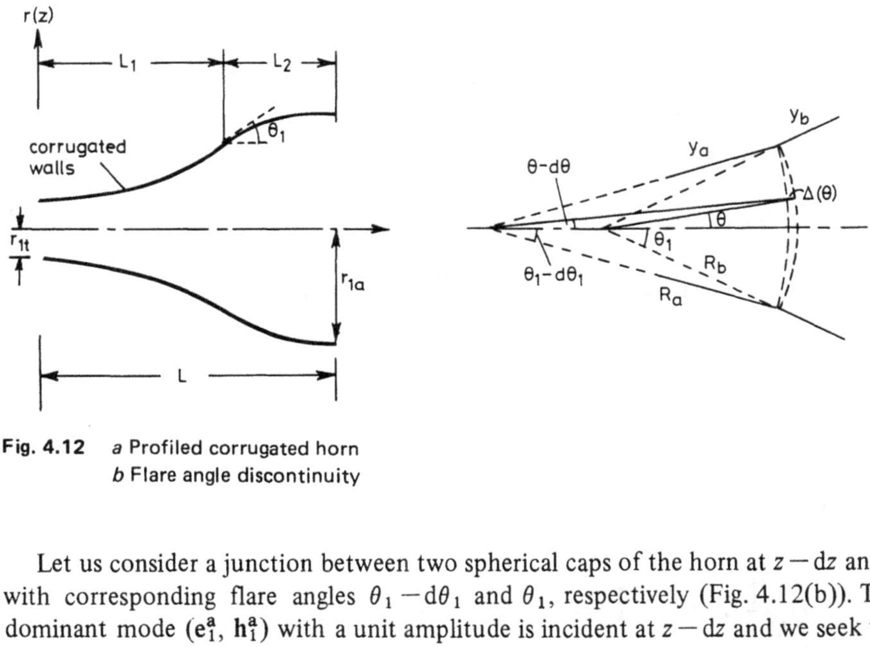

Fig. 4.12 *a* Profiled corrugated horn
b Flare angle discontinuity

Let us consider a junction between two spherical caps of the horn at $z - \mathrm{d}z$ and z with corresponding flare angles $\theta_1 - \mathrm{d}\theta_1$ and θ_1, respectively (Fig. 4.12(b)). The dominant mode $(\mathbf{e}_1^a, \mathbf{h}_1^a)$ with a unit amplitude is incident at $z - \mathrm{d}z$ and we seek the transmitted $\mathrm{d}C_n$, $n > 1$ emerging from z due to mode conversion. Let the modal vector fields at the spherical cap at z be $(\mathbf{e}_n^b, \mathbf{h}_n^b)$, $n = 1, 2 \ldots$.

These modes are the normal modes of a uniform cone having the local geomerical and electrical parameters of the profiled horn. From eqn. 4.57, $\mathrm{d}C_n$ for $n > 1$ is given by

$$dC_n \cong \tfrac{1}{2}[\langle \mathbf{e}_1^a \exp(-j\beta_1^a \Delta(\theta)), \mathbf{h}_n^b\rangle + \langle \mathbf{e}_n^b, \mathbf{h}_1^a \exp(-j\beta_1^a \Delta(\theta))\rangle] - \tfrac{1}{2}[\langle \mathbf{e}_1^b, \mathbf{h}_n^b\rangle + \langle \mathbf{e}_n^b, \mathbf{h}_1^b\rangle] \quad (4.72)$$

where the product $\langle \mathbf{f},\mathbf{g}\rangle$ is defined as the integration of $\mathbf{f}\cdot\mathbf{g}$ over the spherical cap at z. $\Delta(\theta)$ is the radial distance between the two spherical caps at $z-dz$ and z (Fig. 4.12(b)). The second term in eqn. 4.72 arises due to the fact that, in general, the modes are not exactly orthogonal. From the geometry, one can easily prove the following relations:

$$\Delta(\theta) = a(z)\,\mathrm{cosec}^2\theta_1(\cos\theta - \cos\theta_1)\,d\theta_1 \quad (4.73)$$

$$d\theta/d\theta_1 = \sin\theta/\sin\theta_1 \quad (4.74)$$

Next, the θ and ϕ dependence of the $(\mathbf{e}_1^a, \mathbf{h}_1^a)$ at $z-dz$ are displayed explicitly as

$$\mathbf{e}_1^a(\theta - d\theta,\phi) = \mathbf{e}_1^a(\theta,\phi) - d\theta_1\cdot(\partial\mathbf{e}_1^a(\theta,\phi)/\partial\theta)\cdot(\sin\theta/\sin\theta_1) \quad (4.75)$$

and a similar expression holds for $\mathbf{h}_1^a(\theta - d\theta, \phi)$. Furthermore, since $\beta_1^a\Delta(\theta) \ll 1$,

$$\exp(-j\beta_1^a\Delta(\theta) \simeq 1 - j\beta_1^a\Delta(\theta) = 1 - j\beta_1^a(z)\,\mathrm{cosec}^2\theta_1 (\cos\theta - \cos\theta_1)\,d\theta_1 \quad (4.76)$$

Substituting from eqns. 4.75 and 4.76 in eqn. 4.72, dC_n is recast in the form

$$dC_n(z) = Q_1 - Q_2\,\mathrm{cosec}\,\theta_1\,d\theta_1 - Q_3 j\beta_1^a(z)\,\mathrm{cosec}^2\theta_1\,d\theta_1 \quad (4.77)$$

where

$$Q_1 = \tfrac{1}{2}[\langle(\mathbf{e}_1^a - \mathbf{e}_1^b), \mathbf{h}_n^b\rangle + \langle\mathbf{e}_n^b, (\mathbf{h}_1^a - \mathbf{h}_1^b)\rangle] \quad (4.78)$$

$$Q_2 = \tfrac{1}{2}[\langle\sin\theta\,\partial\mathbf{e}_1^a/\partial\theta, \mathbf{h}_n^b\rangle + \langle\mathbf{e}_n^b, \sin\theta\,\partial\mathbf{h}_1^a/\partial\theta\rangle] \quad (4.79)$$

and

$$Q_3 = \tfrac{1}{2}[\langle(\cos\theta - \cos\theta_1)\mathbf{e}_1^a, \mathbf{h}_n^b\rangle + \langle\mathbf{e}_n^b, (\cos\theta - \cos\theta_1)\,\mathbf{h}_1^b\rangle] \quad (4.80)$$

where all fields are calculated at the spherical cap at z with arguments θ and ϕ. The superscripts a and b signify that the modal eigenvalues correspond to the local parameters at $z - dz$ and z, respectively. To continue, resort is made to the explicit field expressions for $(\mathbf{e}_n, \mathbf{h}_n)$ given in eqns. 4.24 and 4.25, which reduce eqns. 4.78–4.80 to

$$Q_1 = A\{L_+(\nu_1^a, \nu_n^b) - L_+(\nu_1^b, \nu_n^b)\} + B\{L_-(\nu_1^a, \nu_n^b) - L_-(\nu_1^b, \nu_n^b)\} \quad (4.81)$$

$$Q_2 = AS_+(\nu_1^a, \nu_n^b) + BS_-(\nu_1^a, \nu_n^b) \quad (4.82)$$

$$Q_3 = AT_+(\nu_1^a, \nu_n^b) + BT_-(\nu_1^a, \nu_n^b) \quad (4.83)$$

where

$$L_{\pm}(\nu, \omega) = \int_0^{\theta_1} f_{\nu\pm}(\theta) f_{\omega\pm}(\theta) \sin\theta d\theta \tag{4.84}$$

$$S_{\pm}(\nu, \omega) = \int_0^{\theta} g_{\nu\pm}(\theta) g_{\omega\pm}(\theta) \sin\theta d\theta \tag{4.85}$$

$$T_{\pm}(\nu, \omega) = \int_0^{\theta} f_{\pm}(\theta) f_{\omega\pm}(\theta) \cos\theta \sin\theta d\theta - \cos\theta_1 L_{\pm}(\nu, \omega) \tag{4.86}$$

$$f_{\nu\pm}(\theta) = \partial P_\nu^1(\cos\theta)/\partial\theta \pm (\sin\theta)^{-1} P_\nu^1(\cos\theta) \tag{4.87}$$

$$g_{\nu\pm}(\theta) = -(\cos\theta \mp 1) f_{\nu\pm}(\theta) - \nu(\nu+1) P_\nu^1(\cos\theta)\sin\theta \tag{4.88}$$

The coefficients A and B are given by

$$\begin{Bmatrix} A \\ B \end{Bmatrix} = \pi/4\,[(\bar{\beta}_1^{\mathrm{a}} + \bar{\beta}_n^{\mathrm{b}})(1 + \bar{\Lambda}_1^{\mathrm{a}}\bar{\Lambda}_1^{\mathrm{b}}) \pm (\bar{\Lambda}_1^{\mathrm{a}} + \bar{\Lambda}_1^{\mathrm{b}})(1 + \bar{\beta}_1^{\mathrm{a}}\bar{\beta}_n^{\mathrm{b}})]/(N_1 N_n)^{1/2} \tag{4.89}$$

$$N_i = \pi/2\,[(\bar{\beta}_i + \bar{\Lambda}_i)(1 + \bar{\beta}_i\bar{\Lambda}_i) L_+(\nu_i, \nu_i) + (\bar{\beta}_i - \bar{\Lambda}_i)(1 - \bar{\beta}_i\bar{\Lambda}_i) L_-(\nu_i, \nu_i)] \tag{4.90}$$

where $\bar{\beta}_i = \beta_i/k$ is the normalised phase constant of mode i relative to the free space wave number, $\bar{\Lambda}_i$ is the normalised modal hybrid factor and N_i is a modal normalisation factor defined previously in eqn. 4.58.

The integrations in eqns. 4.84–4.86 can be obtained in closed forms, then $dC_n(z)$, given by eqn. 4.77, is now completely derived. In order to get some insight into the behaviour of $dC_n(z)$ with the change of the flare angle θ_1, let us consider the simple case of conversion to the HE_{12} mode under the condition $kr(z) \gg 1$ so that both $\bar{\beta}_1$ and $\bar{\beta}_n$ are close to unity. Let us also assume that the balanced hybrid mode condition holds, i.e. the wall admittance $Y = 0$, hence $\bar{\Lambda}_1 = \bar{\Lambda}_n = 1$. With approximations

$$C_n(z)/\partial\theta_1 \simeq \frac{2x_1x_2\,(\theta_1/\sin\theta_1)}{x_1^2 - x_2^2 \sin\theta_1/\theta_1}\,[1 - \mathrm{j}(\theta_1/\sin\theta_1)(3\theta_1 kr/2x_2^2)] \tag{4.91}$$

where $x_1 = 2{\cdot}4048$, $x_2 = 5{\cdot}5201$ are the first two zeros of the Bessel function $J_0(x)$. For typical values of $kr = 10$, $\theta_1 < 25°$, it is seen that $\partial C_n(z)/\partial\theta_1$ is almost a constant. The net conversion level determined by eqn. 4.72 is then highly dependent on the integral phase factor, which, in turn, depends on the horn's profile. Illustrative numerical examples showing this dependence are given below.

The behaviour of $dC_n(z)$ for the EH_{1n} modes, is quite different from that for HE_{1n} modes, namely if the wall admittance $Y = 0$ everywhere, $dC_n(z) = 0$ irrespective of the flare angle variation. This is seen from eqn. 4.89, since then $\bar{\Lambda}_1 = \bar{\Lambda}_n = 1$, hence $A = B = 0$. So, when Y varies around the zero value, $dC_n(z)$ can be shown to be

$$dC_n(z) = -j(K_y dY + K_\theta dY d\theta) \tag{4.92}$$

where dY is the change of the wall admittance (normalised to free space admittance),

$$K_y = \frac{-x_1 x_3 (\sin\theta_1/\theta_1)^{3/2}}{kr(x_1^2 - x_3^2 \sin\theta_1/\theta_1)} \tag{4.93}$$

and

$$K_\theta = \frac{2x_1 (\sin\theta_1/\theta_1)^{1/2}}{kr(x_1^2 - x_3^2 \sin\theta_1/\theta_1)} \left[\frac{x_1^2}{x_1^2 - x_3^2 \sin\theta_1/\theta_1} - j \frac{2kr\,\theta_1^2}{x_3 \sin\theta_1} \right] \tag{4.94}$$

where $x_3 = 5{\cdot}1362$, the first root of $J_2(x)$. The above applies for $kr \gg 1$ and $\theta_1 < 25°$.

Numerical examples

A class of cosine squared profiles has been numerically investigated. Referring to Fig. 4.12(a), the cross-sectional radius $r_1(z)$ is given by

$$r_1(z) - r_{1t} = 2(r_{1a} - r_{1t}) \cos^2\{\pi/2(1 - z/2L_1)\}/(1+\gamma) \ldots 0 < z < L_1$$

$$= \frac{2(r_{1a} - r_{1t})}{1+\gamma} \left[\gamma \cos^2 \left\{ \pi/2 \left(1 - \frac{z + L_2 - L_1}{2L_2}\right)\right\} + \frac{1-\gamma}{2} \right]$$

$$\ldots L_1 < z < L \tag{4.95}$$

where the parameter $\gamma = L_2/L_1$ and determines the location of the maximum flare angle $\hat{\theta}_1$ of the profile which occurs at $z = L_1 = L/(1+\gamma)$ (see Fig. 4.12(a)). The value of $\hat{\theta}_1$ is, however, independent of γ and is given by

$$\tan\hat{\theta} \equiv dr/dz|_{z=L_1} = \pi(r_{1a} - r_{1t})/2L \tag{4.96}$$

$r_1(z)$, $\theta_1(z)$ and $d\theta_1/dz$ are all continuous at $z = L_1$ as can be verified by differentiating eqn. 4.95 once and twice. The case $\gamma = 1$ corresponds to maximum flare angle $\hat{\theta}_1$ half way between the throat and the mouth, while for $\gamma \gtrless 1$, $\hat{\theta}_1$ occurs closer to the throat/mouth, respectively.

HE_{11}-HE_{12} mode conversion

A horn having the profile of eqn. 4.95 with $k_0 r_{1t} = 3{\cdot}0$ and $k_0 r_{1a} = 12{\cdot}0$ is considered. In order to concentrate our study on the effect of flare angle variation, the slot depth is taken constant along the horn. This does not mean, though, that the wall admittance Y is constant, since the later depends also on ka and $\sin\theta_1$. The length of the horn is varied between $k_0 L = 15$ and 40 in Fig. 4.13, with $\gamma = L_2/L_1 = 1{\cdot}0$ and $0{\cdot}4$, and the mode conversion is computed using the above analysis.

As expected, the level of the converted HE_{12} mode at the mouth decreases with increased L. More important is the decreased level of conversion for $\gamma = 0{\cdot}4$ which corresponds to a maximum flare angle $\hat{\theta}_1$ closer to the horn's mouth than to the throat. Explanation for this decrease can be obtained by studying eqn. 4.71. Near

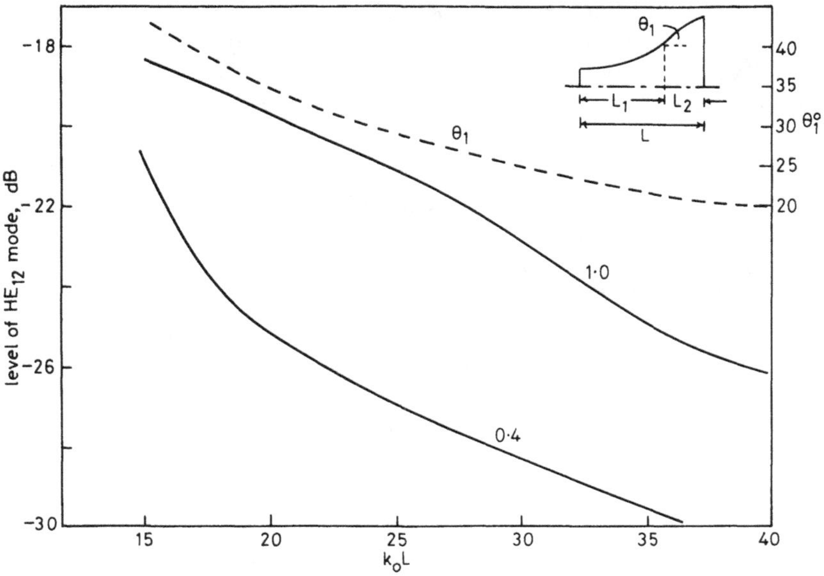

Fig. 4.13 *Level of HE_{12} mode as a function of normalised length of horn for $L_2/L_1 = 1{\cdot}0$ and $0{\cdot}4$. The maximum flare angle θ_1 is also shown as a function of k_0L:$kr_{1t} = 3{\cdot}0$ and $kr_{1a} = 12{\cdot}0$*

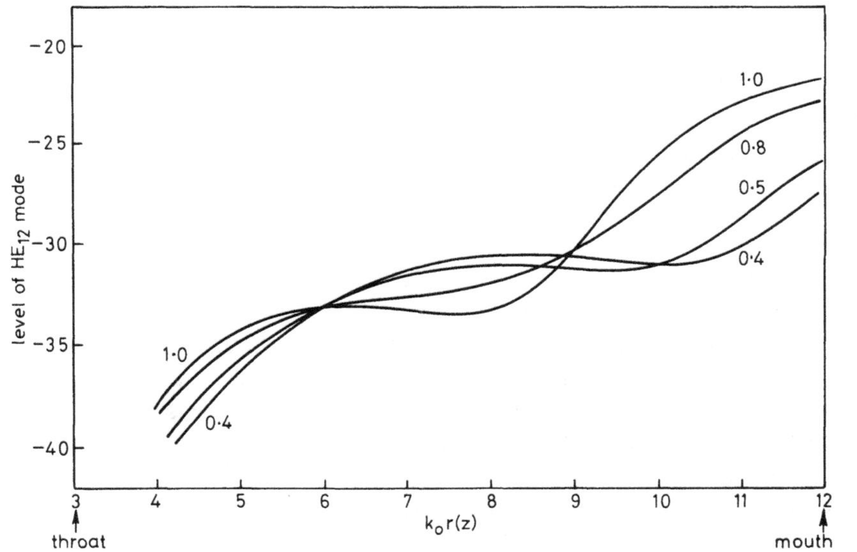

Fig. 4.14 *Level of HE_{12} mode as a function of position along the horn giving by kr_1 for $k_0L = 27.0$. Geometry as Fig. 4.12. Parameter L_2/L_1*

the point $\hat{\theta}_1$ of the profile, $d\theta_1$ changes from positive to negative for $z \gtrless L$, respectively. This means that dC_n changes sign, and, if the phase integral term in eqn. 4.71 is slowly varying, a good deal of cancellation occurs, reducing the overall level of conversion. Now, since $(\beta_1 - \beta_n)$ decreases as $kr_1(z)$ increases, the condition of cancellation is better satisfied as the location of $\hat{\theta}_1$ approaches the horn's mouth, or as γ is reduced. To support this argument, the conversion level of the HE_{12} mode is plotted in Fig. 4.14 as a function of $kr_1(z)$ for $\gamma = 1$, 0·8, 0·5 and 0·4. It is seen that the level of the HE_{12} mode reaches a maximum then drops near the $\hat{\theta}_1$ point. For $\gamma = 1$, this drop is followed by a sharp increase since after this point the differential conversions almost add in phase, as $(\beta_1 - \beta_n)$ becomes small. For $\gamma = 0{\cdot}4$, the process of cancellation extends over a wider region and gets closer to the mouth, hence followed by a relatively smaller increase in HE_{12} power.

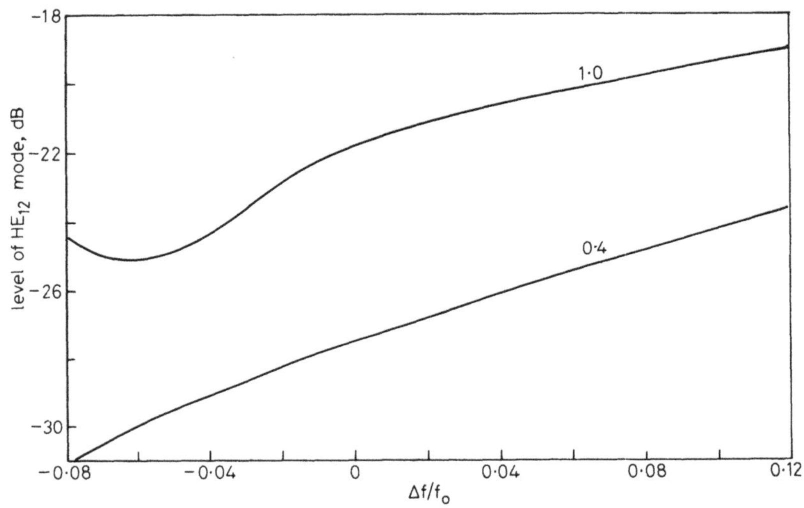

Fig. 4.15 *Level of HE_{12} mode as a function of f/f_0 where f_0 is the frequency where slot depth $= \lambda/4$. Parameters as Fig. 4.14*

Obviously $\hat{\theta}_1$ cannot be located too close to the mouth since then the flare angle will have to be reduced from $\hat{\theta}_1$ to a low value at the mouth in a very small longitudinal distance, which creates a sharp discontinuity. An optimum value for γ in our present example has been found to be $\gamma \cong 0{\cdot}4$. A general conclusion may then be drawn. A profile which has a maximum flare angle in a region of sufficiently high kr_1 (say, $kr_1 > 9$) is likely to have low mode conversion due to phase cancellation occurring in this region. It is to be noted that this cancellation will have a broadband width, since it mainly depends only on the profile shape. This broadband behaviour can be seen in Fig. 4.15, in which the conversion level is shown versus the deviation of the frequency f from f_0 at which the slot is a quarter of a wavelength deep. Generally, the conversion level increases with frequency since the total attenuation of the HE_{12} mode along the horn is reduced with increasing frequency. It is also noted that mode conversion to the HE_{12} mode is mostly attributed to flare angle variations, and to a much lesser extent to changes in wall admittance Y. Actually, for a linear horn of the same length and aperture as for the horn of Fig. 4.15, the mode conversion due to δY has been computed and found to be only of the order of -50 dB or less.

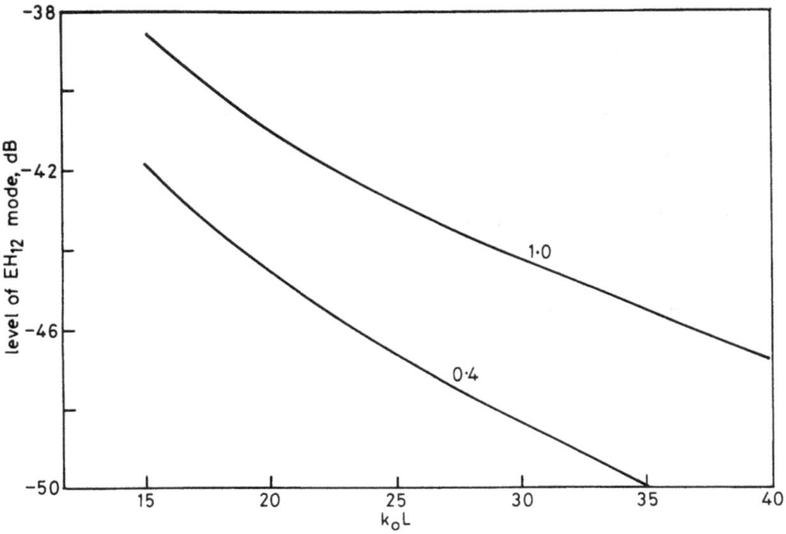

Fig. 4.16 *Level of EH_{12} mode as a function of normalised length of horn for $L_2/L_1 = 1{\cdot}0$ and $0{\cdot}4$. $kr_{1t} = 3{\cdot}0$ and $kr_{1a} = 12{\cdot}0$*

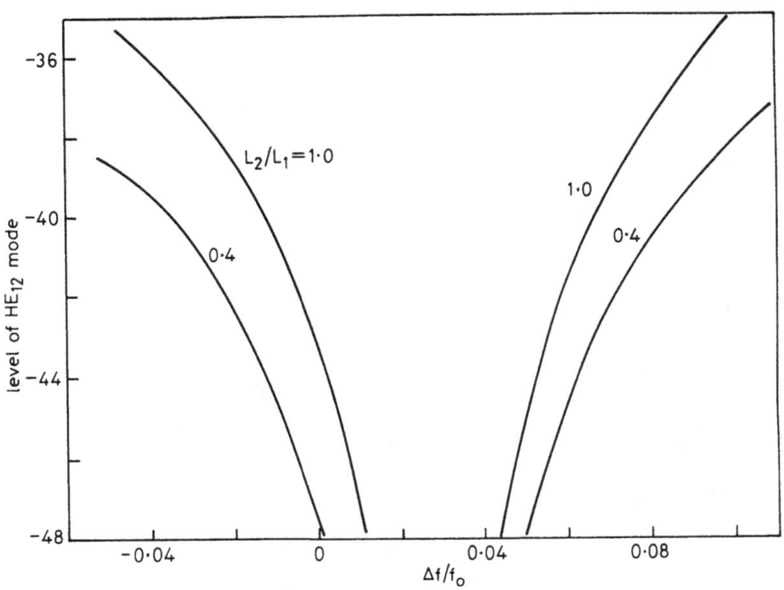

Fig. 4.17 *Level of EH_{12} mode as a function of f/f_0. Parameters as in Fig. 4.14*

HE_{11}-EH_{12} mode conversion

Results for conversion levels of the EH_{12} mode for the same class of horn profiles are given in Figs. 4.16 and 4.17. The behaviour of the level of conversion with k_0L and $\gamma = L_2/L_1$ in Fig. 4.16 is generally the same as in the case of HE_{12} mode, but here, considerably lower conversion levels occur. This may be attributed to the heavy dependence on δY for the EH_{12} mode. This dependence is manifested in Fig. 4.17, where a deep minimum of the conversion level occurs near the centre frequency where δY is very small.

4.5 Radiation from corrugated horns by the Kirchhoff–Huygen method

The aperture fields of a conical corrugated horn are given in eqns. 4.2 *a–d*. The radiation from the spherical constant-phase surface in the aperture can be obtained directly by means of a Kirchhoff–Huygen integration. Following Clarricoats and Saha [30] we express the radiation field at P, $\mathbf{E_p}(R', \theta', \phi')$ as

$$\mathbf{E_p}(R', \theta', \phi') = \frac{\mathrm{j}k \exp(-\mathrm{j}kR')}{4\pi R'} \hat{i}_{R'} \times \int_s \{\hat{i}_n \times \mathbf{E_{tang}}$$

$$- z_0 \times (i_n \times \mathbf{H_{tang}}) \exp(\mathrm{j}k\mathbf{R}_0 \cdot i_{R'} \mathrm{d}s) \tag{4.97}$$

$\hat{i}_{R'}$, $\hat{i}_n$ are unit vectors and the fields E_{tang}, H_{tang} are tangential to the surface of the spherical cap. On replacing the factor outside the integral by c, we obtain, after some algebraic manipulation:

$$E_{p\theta}(\theta', \phi') = c \int_0^{2\pi} \int_0^{\theta_1} [-E_\theta \cos(\phi' - \phi) - E_\phi \cos\theta \sin(\phi' - \phi)$$

$$+ z_0 H_\theta \cos\theta' \sin(\phi' - \phi) - z_0 H\phi \{\sin\theta \sin\theta' + \cos\theta \cos\theta' \cos$$

$$(\phi' - \phi)\}] \exp[\mathrm{j}kR_0\{\cos\theta \cos\theta' + \sin\theta \sin\theta' \cos(\phi' - \phi)\}]$$

$$\sin\theta \,\mathrm{d}\theta \,\mathrm{d}\phi \tag{4.98}$$

$$E_{p\phi}(\theta', \phi') = c \int_0^{\pi} \int_0^{\theta_1} [E_\theta \cos\theta' \sin(\phi' - \phi) - E_\phi \{\sin\theta \sin\theta'$$

$$+ \cos\theta \cos\theta' \cos(\theta' - \phi)\} + z_0 H_\theta \cos(\phi' - \phi)$$

$$+ z_0 H_\phi \cos\theta \sin(\phi' - \phi)] \exp[\mathrm{j}kR_0\{\cos\theta \cos\theta'$$

$$+ \sin\theta \sin\theta' \cos(\phi' - \phi)\}] \sin\theta \,\mathrm{d}\theta \,\mathrm{d}\phi \tag{4.99}$$

Let $a = kR_0 \cos\theta'$ and $b = kR_0 \sin\theta'$, and let

$$E_\theta(\theta, \phi) = -g(\theta) C \cos m\phi \tag{4.100}$$

$$E_\phi(\theta, \phi) = h(\theta) C \sin m\phi \tag{4.101}$$

On substituting for E_θ and E_ϕ, we obtain, after performing the integration and combining the constants,

$$E_{p\theta}(\theta', \phi') = D \cos m\phi' \int_0^{\theta_1} [2jg(\theta) J_m(b \sin\theta) \sin\theta \sin\theta'$$

$$+ (j)^{m+1} J_{m+1}(b \sin\theta) \{g(\theta)(1 + \cos\theta \cos\theta')$$

$$\mp h(\theta)(\cos\theta + \cos\theta')\} + j^{m-1} J_{m-1}(b \sin\theta)$$

$$g(\theta)(1 + \cos\theta \cos\theta') \pm h(\theta)(\cos\theta + \cos\theta')\}]$$

$$e^{(ja\cos\theta)} \sin\theta d\theta \qquad (4.102)$$

The integral for $E_{p\phi}(\theta', \phi')$ is similar to that for $E_{p\theta}$ except that $g(\theta)$ and $h(\theta)$ are interchanged and $\cos m\phi'$ is replaced by $-\sin m\phi'$. When $\bar{\Lambda} = +1$, we have

$$g(\theta) = h(\theta) = f_{\nu+}^m(\theta) = \frac{mP_\nu^m(\cos\theta)}{\sin\theta} + \frac{dP_\nu^m(\cos\theta)}{d\theta} \qquad (4.103)$$

When $\bar{\Lambda} = -1$, we have

$$g(\theta) = h(\theta) = f_{\nu-}^m(\theta) = \frac{dP_\nu^m(\cos\theta)}{d\theta} - \frac{mP_\nu^m(\cos\theta)}{\sin\theta} \qquad (4.104)$$

For $m = 1, \bar{\Lambda} = +1$

$$E_{p\theta \atop \phi} = \pm F_r(\theta') + jF_i(\theta') {\cos \atop \sin} \phi' \qquad (4.105)$$

$$F_{r \atop i}(\theta') = D\int_0^{\theta_1} f_{\nu+}^1(\theta) G_{r \atop i}(\theta, \theta') d\theta$$

and

$$G_{r \atop i}(\theta, \theta') = (1 + \cos\theta)(1 + \cos\theta') \sin\theta {\cos \atop \sin} (a \cos\theta) J_0(b \sin\theta)\}$$

$$- \{(1 - \cos\theta)(1 - \cos\theta') \sin\theta {\cos \atop \sin} (a\cos\theta) J_2(b \sin\theta)\}$$

$$\mp \{2 \sin\theta' \sin^2\theta {\sin \atop \cos} (a \cos\theta) J_1(b \sin\theta)\}$$

From the above equation, we observe that, under balanced hybrid conditions, pattern symmetry exists about the axis $\theta' = 0$. A maximum in the boresight direction $\theta' = 0$ occurs for HE_{1n} modes, while for EH_{1n} modes there is a null on the boresight, and similarly for all modes with $m > 1$. There is no cross-polarised component of radiated field if the aperture field is linearly polarised.

Comparison between Kirchhoff–Huygen and SWEX methods
In Fig. 4.18, we compare copolar radiation patterns predicted by the Kirchhoff–Huygen and modal expansion methods for $\theta_1 = 50°, 60°$ and $70°$; the agreement is excellent. We have studied the convergence of the series and have found that for various observation angles, the pattern converges to within 1% when $N \simeq kR_0$. A good approximation to the final pattern is also obtained with only about six terms

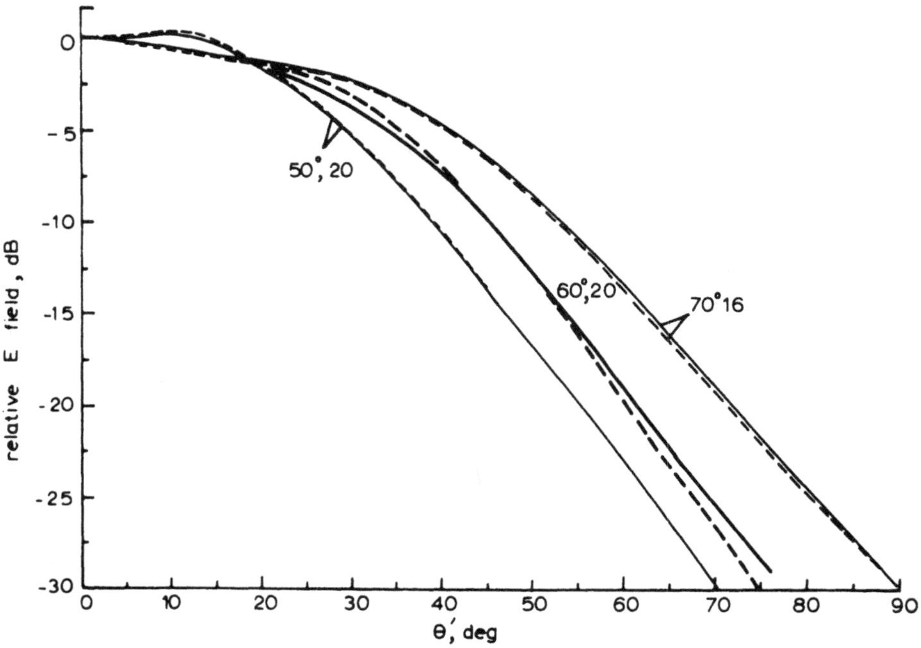

Fig. 4.18 *E plane radiation patterns for wide angle corrugated horns*
——— modal expansion method
– – – – Kirchhoff–Huygen method Parameters θ_1 and kR_0

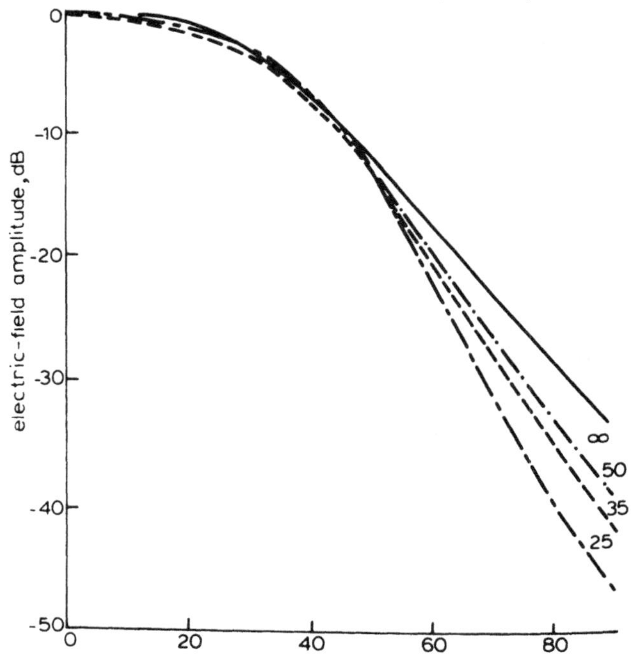

Fig. 4.19 *Near-field of conical corrugated horn. Parameter normalised distance kR from horn apex. $kR_0 = 20$*

in the series, especially for wide flare-angle horns. It is also possible to use the above method to determine the near-field of corrugated horns as shown in Fig. 4.19, computed using spherical wave expansion methods, (see also Chapter 5).

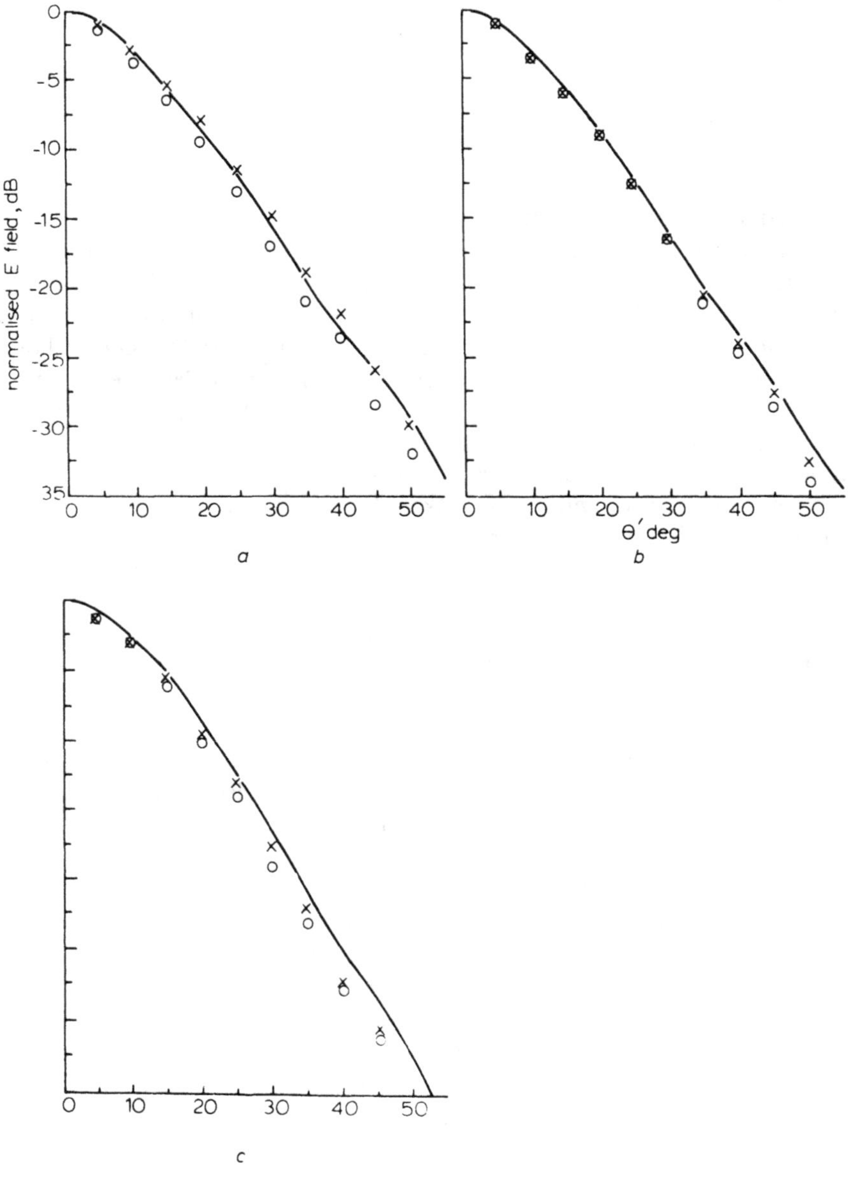

Fig. 4.20 *Radiation patterns of 30° semi-flare angle corrugated horn*
× H plane experimental ○ E plane experimental points ——— Theory using spherical wave expansion method *a* 8·5 kHz *b* 9·0 kHz *c* 11·0 kHz

4.6 Comparison of predicted and measured radiation patterns

The theory described in the preceeding sections has now been widely validated for corrugated waveguides and horns, the latter with θ_1 up to about 75°. Results for a corrugated horn with 30° semi-flare angle are shown in Fig. 4.20, and further discussion of experimental results are to be found in Chapter 5.

4.7 Radiation from corrugated horns by cylindrical waveguide approximation

As we have seen in previous sections the copolar and crosspolar radiation patterns of conical corrugated horns can be predicted using spherical wave expansion and Kirchhoff–Huygen methods. However, this is necessarily approximate because the analysis must assume that the surface admittance of the corrugations is independent of position along the horn. Only under this condition is the spherical wave equation in a conical co-ordinate system separable into independent parts. A second approximation exists for the wave admittance $-H_\phi/E_R$ at the boundary $\theta = \theta_1$ precluding a study of the slot geometry.

None of these assumptions is serious as far as the copolar radiation pattern prediction is concerned, and many workers have obtained good agreement with experimental measurements. The prediction of the crosspolar radiation pattern for narrow flare-angle horns requires a more precise theoretical analysis because the geometry of the slots is a crucial parameter in determining the peak cross-polar power level. The analysis of an open-ended corrugated waveguide, including space harmonics, enables the influence of slot geometry to be studied. The horn aperture is treated as an open-ended corrugated waveguide and the waveguide aperture fields computed. A simple spherical phase factor, dependent on the horn flare angle, is then added to the aperture fields and the radiation pattern computed. The method is valid certainly up to flare angles of 20° and even for larger angles the method gives a good approximation to the measured patterns.

We show that this approach is justified by comparing the characteristic equation and aperture fields of narrow flare-angle horns and waveguides. We start by showing that the aperture fields, expressed in cylindrical co-ordinates, for a narrow flare-angle conical corrugated horn are identical to the aperture fields for a corrugated circular waveguide with the addition of a spherical phase factor.

Consider the case when the slot depth is such that the wave admittance, H_ϕ/E_R at the mouth of the slots is zero; this is the balanced-hybrid condition. The characteristic equation for the spherical-hybrid mode in a narrow flare-angle conical corrugated horn with a small semi flare angle θ_1, and a large normalised slant length, kR_0, is given by

$$\alpha\theta \frac{J_1'(\alpha\theta_1)}{J_1(\alpha\theta_1)} = -1 - \left[\frac{\alpha\theta_1}{kr_1}\right]^2 \qquad (4.106)$$

where $R\theta_1 = r_1$, $\alpha = \sqrt{\{\nu(\nu+1)\}}$, ν is the order of the Legendre function, and the Bessel function approximation to the Legendre function has been used. For comparison the characteristic equation for the corrugated circular waveguide, radius r_1, is given by

$$(Kr_1)\frac{J_1'(Kr_1)}{J_1(Kr_1)} = -1 - \left[\frac{Kr_1}{kr_1}\right]^2 = -\bar{\beta} \tag{4.107}$$

Hence eqns. 4.106 and 4.107 are identical when $\alpha\theta_1$ is equated to Kr_1, implying identical wavenumbers.

We now consider the horn aperture fields at the balanced-hybrid condition. The aperture electric fields defined over a spherical cap for the corrugated conical horn for small θ_1 and large kR are given by

$$\mathbf{E}_\theta = \alpha_1\left[(k)\frac{J_1(\alpha\theta_1)}{r} + k\left(1-\left[\frac{\alpha\theta_1}{kr_1}\right]^2\right)J_1'(\alpha\theta_1)\frac{\alpha\theta_1}{r}\right]\hat{i}_\theta e^{-jkR} \tag{4.108}$$

$$\mathbf{E}_\phi = \alpha_1\left[(k)\left(\frac{\alpha\theta_1}{r}\right)J_1'(\alpha\theta_0) + k\left(1-\left[\frac{\alpha\theta_1}{kr_1}\right]^2\right)\frac{J_1(\alpha\theta_1)}{r}\right]\hat{i}_\phi e^{-jkR} \tag{4.109}$$

The transform from spherical co-ordinates (R, θ, ϕ) to cylindrical co-ordinates (r, ϕ, z) is given by $\hat{i}_R = \hat{i}_r \sin\theta + \hat{i}_z \cos\theta$, $\hat{i}_\theta = \hat{i}_r \cos\theta - \hat{i}_z \sin\theta$ and $\hat{i}_\phi = \hat{i}_\phi$. When θ is in the range $0 \leqslant |\theta| \leqslant \theta_1$ and θ_1 is small, eqns. 4.108 and 4.109 become

$$\begin{aligned}\mathbf{E}_r &\propto E_r'\hat{i}_r e^{-jkz}\\ \mathbf{E}_\phi &\propto E_\phi'\hat{i}_\phi e^{-jkz}\end{aligned} \tag{4.110}$$

Expressing the phase factor in terms of the radial r component in the aperture, Fig. 4.21 yields:

$$e^{-jkz} = e^{-jk(z_1+\Delta)} \tag{4.111}$$

By moving the phase reference to the plane aperture $z = z_1$, and expressing Δ in terms of r, the phase factor becomes

$$\exp\left[-jkR\left(1-\sqrt{1-\left(\frac{r}{R}\right)^2}\right)\right] \equiv e^{j\phi(r)} \tag{4.112}$$

Hence, the aperture electric fields for the narrow flare angle conical corrugated horn, expressed in cylindrical co-ordinates, are given by

$$\mathbf{E}_r \propto \left[(k)\frac{J_1(\alpha\theta_1)}{r} + k\left(1-\left[\frac{\alpha\theta_1}{kr_1}\right]^2\right)J_1'(\alpha\theta_1)\frac{\alpha\theta_1}{r}\right]\hat{i}_r e^{j\phi(r)} \tag{4.113}$$

$$\mathbf{E}_\phi \propto \left[(k)\left(\frac{\alpha\theta_1}{r}\right)J_1'(\alpha\theta_1) + k\left(\left[\frac{\alpha\theta_1}{kr_1}\right]^2\right)\frac{J_1(\alpha\theta_1)}{r}\right]\hat{i}_\phi e^{j\phi(r)} \tag{4.114}$$

Comparison of these equations with the aperture electric fields for the corrugated circular waveguide shows them to be identical for $\alpha\theta_1 \equiv Kr_1$, if they are multiplied by the spherical phase factor $\phi(r)$.

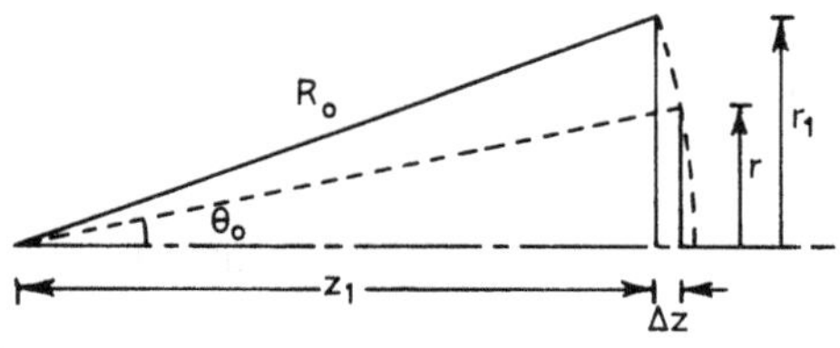

Fig. 4.21 *Geometry for spherical phase cap*

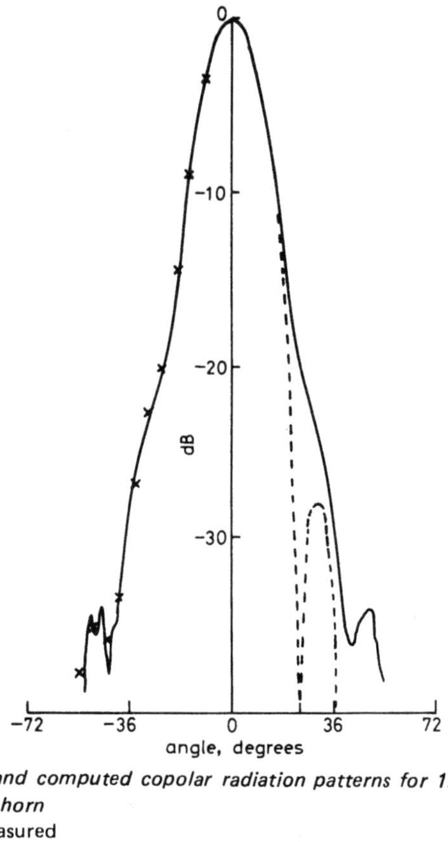

Fig. 4.22 *Measured and computed copolar radiation patterns for 12 degree semi-flare angle corrugated horn*

——— measured

– – – – computed using waveguide approximation

x x x x computed with spherical phase cap

$r_1 = 63{\cdot}5$ mm, $s = 9$ mm, $b = 7$ mm, $t/b = 2{\cdot}5$ mm

This analysis is also valid for the more exact space harmonic analysis where the fields in the waveguide are represented as a series of space harmonics, and the fields in the corrugations as a series of evanescent radial modes. The method is more

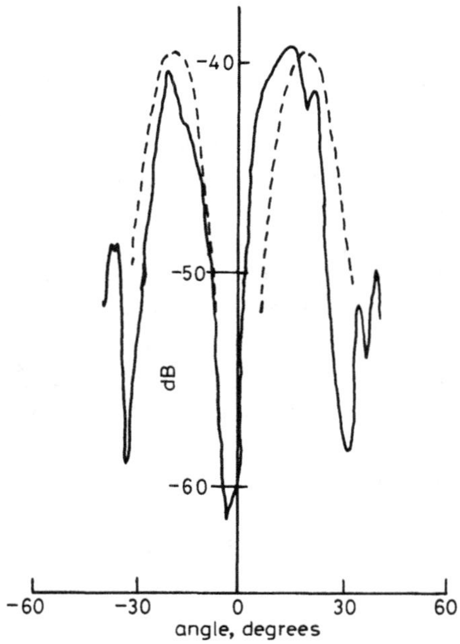

Fig. 4.23 *Measured and theoretical ($K = L = 1$) (see Section 3.3) crosspolar patterns at 10 GHz for horn of Fig. 4.22*

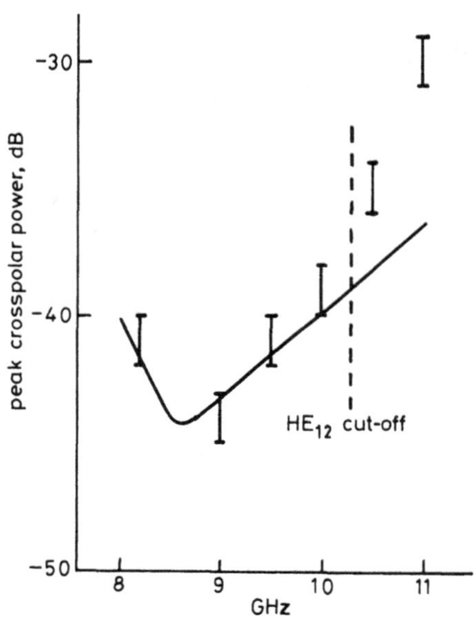

Fig. 4.24 *Measured and theoretical ($K = L = 1$) (see Section 3.3) peak crosspolarisation for horn of Fig. 4.22*

complicated than the surface admittance method but it does allow the exact profile of the corrugations to be specified.

Thus the radiation pattern of a narrow flare-angle conical corrugated horn can be represented by expressing the aperture fields as those of the HE_{11} mode (including space harmonics) of the equivalent waveguide, with the addition of a spherical phase factor. The far-field radiation pattern can then be calculated using an aperture field method.

Fig. 4.22 compares the measured copolar radiation patterns for a 12° semi flare-angle corrugated horn with theoretical points predicted using the waveguide approximation described here. Excellent agreement is obtained using the waveguide approximation, with the first sidelobe of the 12° horn being accurately predicted.

Fig. 4.23 compares the measured crosspolar radiation pattern for the 12° semi flare angle corrugated horn with the theoretical prediction using the method described here. The peaks of the measured cross polar patterns plotted as a function of frequency are shown in Fig. 4.24 and good agreement between experiment and theory is obtained up to a frequency of 10·4 GHz. Above this frequency the measured pattern becomes higher than that theoretically predicted and this is attributed to coupling into the HE_{12} mode.

4.8 Radiation from conical corrugated horns by means of the Laguerre–Gaussian expansion method

Although the spherical-wave expansion method provides a means to predict accurately the near-field and far-field of a conical corrugated horn, when the semi flare angle is small and the horn length large, in terms of wavelengths, many terms are required in the expansion. With present generation mainframe computers this presents little problem but there is attraction in a method which potentially brings the prediction method for copolar patterns within the range of smaller computers. Such a method is the Laguerre–Gaussian expansion investigated by Bitter and Aubry [17]. The method offers another advantage for millimetre wave antenna designers using Gaussian optics since the first term in the series expansion corresponds to a Gaussian beam mode.

Under balanced hybrid conditions we have for the Cartesian component of the aperture field E_x in a horn of length L

$$E_x(r,\theta) = J_m(Kr)\exp\left(-j\frac{kr^2}{2L}\right)\cos m\phi \tag{4.115}$$

We note that $m = 0$, corresponds to the HE_{11} mode, $m = 1$, to the TM_{01} mode. Next expand E_x as a Laguerre–Gaussian series

$$E_x(r,\theta) = \exp\left(-j\frac{kr^2}{2L}\right)\sum_{n=0}^{\infty} A_n^m f_n^m(r,\theta) \tag{4.116}$$

where

$$f_n^m(r,\theta) = \left(\frac{\sqrt{2}r}{W}\right)^m L_n^m\left(\frac{2r^2}{W^2}\right)\exp\left(\frac{-r^2}{W^2}\right)\cos m\phi$$

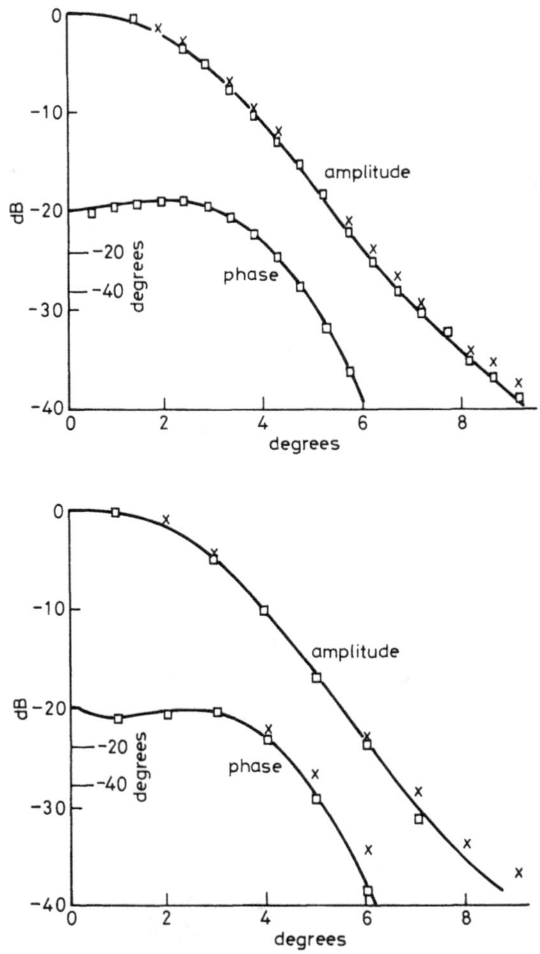

Fig. 4.25 *Radiation pattern computed using Laguerre-Gaussian theory [17]*
———— theoretical ($N = 9$), $\theta_1 = 5.8$ $R_0 = 2487$ mm, Distance in front of aperture = 1300 mm ○ ○ ○ ○ H plane measured x x x x E plane measured *a* 11·1 GHz *b* 14·25 GHz

L_n^m is the generalised Laguerre polynomial of order n and m and W is a parameter defining the so-called beam waist of the Gaussian beam. Note that the first term of the series with $m = 0$ (HE_{11} mode) is the Gaussian function, while with $m = 1$ (TE_{01} mode) the first term is the derivative of the Gaussian function.

To proceed it is first necessary to select a value for W which maximises the quantity η where

$$\eta = \sum_{n=0}^{N} \frac{|A_n^m|^2 |f_n^m|^2}{|J_m^T|^2} \tag{4.117}$$

and J_m^T is the function $J_m(Kr), r < r_1$.

To obtain the radiation field in cylindrical co-ordinates referred to the horn aperture, the Kirchhoff–Huygens integration over the horn aperture yields

$$E_p(\rho, \phi, z) = \sum_{n=0}^{N=1} A_n B \exp(-jkz) \left(\frac{B}{B_r}\right)^n L_n \left(|B|^2 \frac{2\rho^2}{W^2}\right)$$
$$\exp[-jk(1-B)\rho^2/2z] \tag{4.118}$$

where

$$B = \frac{jk}{2\gamma z} \quad \gamma = \frac{1}{W^2} + jk\frac{L+z}{2Lz}$$

L = length of horn.

A useful simplification can be made by taking only one term in the series, thus for the HE_{11} mode

$$J_0^T(Kr) = A_0 f_0(r, \theta) \exp\left(-j\frac{kr^2}{2L}\right) \tag{4.119}$$

Experience also suggests that $W = 0{\cdot}64 r_1$ optimises eqn. 4.117. Physically the radiation behaviour of the corrugated horn is approximated by a Gaussian beam with beam waist W just less than 2/3 of the aperture radius.

Bitter and Aubry have applied the Laguerre–Gaussian method and have compared predicted copolar patterns with those obtained experimentally. Fig. 4.25 compares far-field amplitude and phase measured for a 5·8° semi flare-angle horn with predictions with $N = 9$. Clearly the agreement is excellent. Obviously the patterns could be obtained by the other methods described in Sections 4.3, 4.5 and 4.6. However, we should note that for accuracy with the spherical wave expansion, a very large number of terms would be required. Were that number, taken to be equal to kL in the example of Fig. 4.25, 581 terms would be required, which is obviously numerically unacceptable.

Chapter 5

Design of cylindrical and conical corrugated horns

5.1 Introduction

This chapter is concerned with the factors which influence the design of conical corrugated horns. We start by surveying these factors, then study the electrical specifications and the procedures needed to design the horn to satisfy the specifications. A large number of design curves are included. The information needed to design various types of corrugated horns is summarised. Finally computer prediction is discussed.

There are a wide variety of antenna applications where a corrugated horn is used. Although it is occasionally used by itself, for instance as a gain standard, it is usually used as a feed for a reflector antenna. The desirable characteristics, which have been discussed in the previous chapters, all make the corrugated horn an ideal candidate for use as a feed. These uses can be divided into two types, prime focus reflectors and multireflector antennas, the Cassegrain and the Gregorian reflectors being the main uses, though sometimes more than two reflectors are used as in the beam waveguide feed system for large earth stations of the INTELSAT-A class. Millimeter wave radio astronomy antennas also often use more than two reflectors. The increasing sophistication of spacecraft antennas is one area where new designs have evolved, such as the compact profiled horns and multifrequency horns. The types of corrugated horn together with typical uses are listed in Table 5.1. They will be dealt with in more detail in Section 5.8. The reason for identifying the different types is that each type has common characteristics and problems as far as the designer is concerned. The larger aperture narrow flare angle horns, which form the largest single class of horns built so far, are the easiest to design because the large diameter makes the horn much less sensitive to crosspolarisation and mode conversion problems. The small aperture horns have a performance which is strictly limited by the aperture diameter. The designer must take into account the effects of the flange which is inevitably present at the aperture of the horn. The bandwidth of this type of horn is often a problem. The wide flare angle horns (sometimes referred to as scalar feeds) are potentially attractive because the copolar radiation pattern depends on angle rather than aperture size, they can have a very large

Table 5.1 *Types of corrugated horns and their uses*

Types of horn	Typical uses
Large aperture, narrow flare angle $\left(\frac{D}{\lambda} > 4, \theta_1 < 15^\circ\right)$	Large Cassegrain antennas Earth stations Radio astronomy antennas
Small aperture, narrow flare angle $\left(\frac{D}{\lambda} < 4, \theta_1 < 15^\circ\right)$	Spacecraft antennas Prime focus
Wide flare angle $(\theta_1 > 15^\circ)$	Prime focus reflectors Gain standard Low frequency
Multimode horns	Radars Prime focus reflectors Offset reflectors
Very broad band horns	Electronic warfare
Multifrequency	Multifrequency communication channels
Compact horns	Spacecraft antennas

bandwidth with nearly constant copolar radiation characteristics. We have found that they can also have very low crosspolarisation characteristics. However the beamwidth of the patterns produced by the wide flare angle horns limits the range of applications. The remaining types of horn listed in Table 5.1 are specialist types which are derivatives from the first three types. The characteristic of the multimode tracking horns, as the name implies, is the ability to propagate and radiate higher order modes in addition to the fundamental mode. Very broad band horns are used for situations where a number of frequency channels are required to be received, or radiated by the one horn. The crosspolar characteristics and the match at the throat of the horn are basically narrow band, so broad band horns can only be obtained by some compromise of the characteristics. A related type of horn is the multifrequency horn where the aim is to pass two or more separate signal channels with good characteristics at each channel frequency. Compact horns are horns where the taper from throat to aperture is nonlinear, producing a shorter horn than with a linear taper but retaining the desirable radiation characteristics of a narrow flare angle horn.

The electrical parameters which determine the performance of the corrugated

Table 5.2 *Design parameters for corrugated horns*

Radiation patterns
Principal plane patterns
Copolar patterns
Pattern symmetry (E and H plane equality)
Beamwidths
Crosspolar characteristics
Frequency characteristics
Frequency of operation
Bandwidth
Multiband operation
Efficiency as a feed for a reflector
Total aperture illumination
Spillover
Phase
Crosspolar
G/T for a reflector
Gain
Input impedance and VSWR
Phase centre and movement
Differential phase shift along horn
Multimode operation
Mechanical
External diameter
Length
Weight
Ease of manufacture

horn are listed in Table 5.2 not all the parameters are of interest in one application. Often one or two parameters are of overriding concern and will determine the way the horn is designed and manufactured. These parameters will be referred to throughout the remainder of this chapter. Most are self-explanatory or will be defined when they appear in the text.

In any type of corrugated horn we can identify four parts of the horn for design purposes, since to some extent each part can be considered separately. The four parts are shown in Fig. 5.1. They are (i) the aperture diameter and flare angle which principally determine the copolar beamwidth; (ii) the corrugations, which determine the pattern symmetry and the crosspolar characteristics; (iii) the flare section between the throat and the aperture which determines the position of the phase centre and the generation of any higher order modes along the horn; (iv) the throat region which determines both the impedance match into the section of waveguide behind the horn and the mode conversion level at the throat. Each of these parts will be investigated in detail in the next four sections. Where theoretical results are presented they have been computed using as exact a model as is necessary for the problem. Copolar patterns can be adequately predicted with a simple representation of the horn aperture fields, but crosspolar patterns need a more exact model which includes the corrugation geometry if accurate design data is desired. The theoretical methods have been studied in detail in the preceeding chapters. Therefore, in this chapter the background theory will be assumed and only the results and explanation presented.

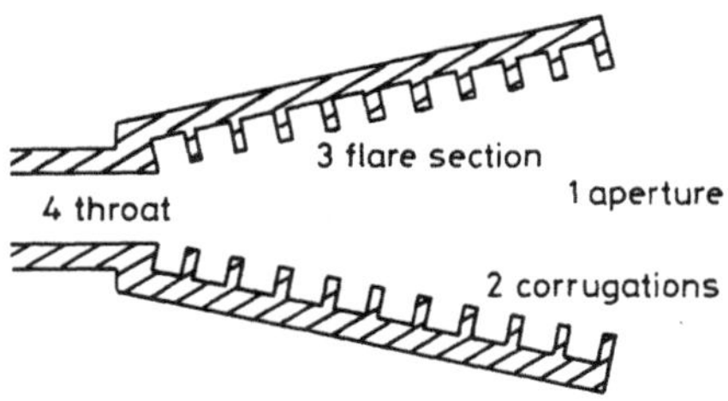

Fig. 5.1 *Corrugated horn showing regions relevant for design*

5.2 Copolar Radiation Characteristics

The copolar radiation characteristics of a conical corrugated feed are determined mainly by the aperture diameter and the horn flare angle. We shall describe here

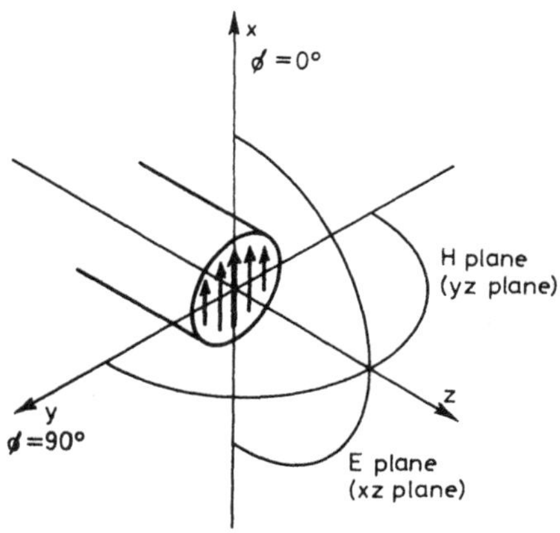

Fig. 5.2 *Planes of polarisation*

how they, and other factors, influence the shape of the radiation patterns. The term *copolar* will be taken to include the radiation pattern of the horn in any plane, angle ϕ with respect to the reference axis, Fig. 5.2, where the field is parallel to the field of the source. The *crosspolar* field is the orthogonal component. The designer is usually most interested in the principal planes, i.e. the $E\,(\phi = 0^\circ)$ and $H\,(\phi = 90^\circ)$ planes and the $\phi = 45^\circ$ plane where the crosspolarisation is maximum. Because of the importance of being clear about the definition of polarisation it is worth restating the definition universally accepted for horn and feed studies. This is the so-called Ludwig Third definition [185] which states that the reference polarisation is that of a Huygens source (electric and magnetic dipoles with orthogonal axes lying in the aperture plane and radiating equal fields in phase along the z-axis). The crosspolarisation is then the polarisation of a similar source rotated 90° in the aperture plane. This definition corresponds most closely to what is normally measured on the antenna range and interchanging the copolar and crosspolar fields as measured in any direction, corresponds to a 90° rotation of the reference source.

Expressed mathematically, the copolar $E_p(\theta, \phi)$ and crosspolar $E_q(\theta, \phi)$ fields are given in terms of the spherical co-ordinate components $E_\theta(\theta, \phi)$ and $E_\phi(\theta, \phi)$ by

$$\begin{bmatrix} E_p(\theta,\phi) \\ E_q(\theta,\phi) \end{bmatrix} = \begin{bmatrix} \cos\phi & -\sin\phi \\ \sin\phi & \cos\phi \end{bmatrix} \begin{bmatrix} E_\theta(\theta,\phi) \\ E_\phi(\theta,\phi) \end{bmatrix} \tag{5.1}$$

where it is assumed that the principal electric field vector is aligned with the x-axis, Fig. 5.2. If the electric field vector is aligned with the y-axis, eqn. 5.1 still applies but the copolar compoent is given by E_q and the crosspolar component by E_p.

The transformation from the aperture fields of a conical horn to the far-field gives $E_\theta(\theta, \phi) = C_\theta(\theta) \cos m\phi$ and $E_\phi(\theta, \phi) = C_\phi(\theta) \sin m\phi$, where $C_\theta(\theta)$ and $C_\phi(\theta)$ are the pattern functions. Combining these relations with eqn. 5.1 for a unity azimuthally dependent mode, gives

$$\begin{aligned} E_p(\theta,\phi) &= C_\theta(\theta)\cos^2\phi + C_\phi(\theta)\sin^2\phi \\ E_q(\theta,\phi) &= \tfrac{1}{2}\left[C_\theta(\theta) - C_\phi(\theta)\right]\sin 2\phi \end{aligned} \tag{5.2}$$

The crosspolar field is a direct measure of the difference between the E plane ($\phi = 0°$) and H plane ($\phi = 90°$) copolar patterns and is a maximum in the $\phi = 45°$ and 135° plane.

Radiation patterns are usually plotted as E_p and E_q (dB amplitude) against the θ angular co-ordinate for the $\phi = 0$, 45 and 90° planes.

The availability of fast computers with graphical output has increased the use of contour plots, see Fig. 5.3, which displays the complete copolar or crosspolar pattern. However, for ease of explanation the rectangular decibel plot is preferred and will be largely used in the following sections.

We shall assume initially that the HE_{11} mode alone is propagating in the horn, the patterns and influence of higher order mode distortion will be dealt with later in the section. As is well known, the copolar radiation pattern of a corrugated horn or waveguide operating under balanced hybrid conditions is basically Gaussian in shape, at least down to about the – 15 dB power level. This fact is shown in Fig. 5.4, which compares the pattern of an actual corrugated waveguide to an ideal Gaussian pattern. This and subsequent theoretical computed patterns for the corrugated horn have been obtained using the exact space-harmonic model unless stated otherwise. Also, if only one copolar pattern is plotted this implies that the horn radiates a low level of crosspolarisation and there is negligible difference between the principal plane patterns.

In considering copolar patterns it should be remembered that we are principally interested in the use of corrugated horns as feeds for reflectors. This means that the copolar pattern is illuminating the reflector out to the edge angle of the reflector.

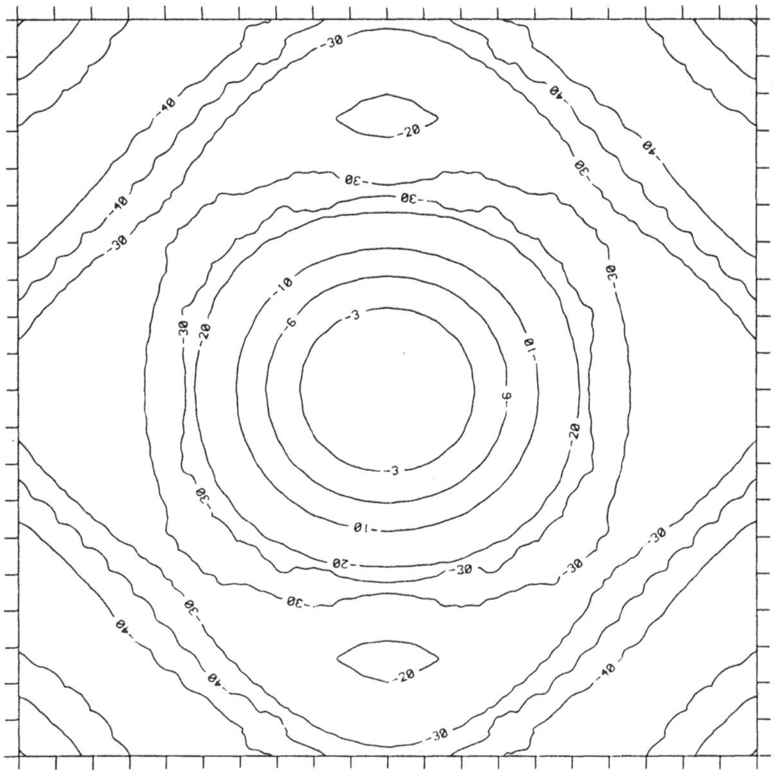

Fig. 5.3 *Contour plot of copolar radiation pattern*

This angle normally corresponds to a pattern level of about -10 dB to give the optimum efficiency (see Section 5.7) but in some low spillover designs an edge taper of -20 dB may be required. The patterns at angles outside this level are rarely of interest, except in so far as the energy at these angles reduces the overall efficiency. As a consequence, we can give portions of the pattern outside the main beam only minor consideration, and note the justification for treating the corrugated horn as a Gaussian pattern radiator. This fact is particularly useful as it opens up the possibility of treating the radiation from the horn as a Gaussian beam. Multiple reflector feed systems have been designed in this way, for instance the beam waveguide feed used in many INTELSAT Type A earth stations and the Nasmyth and Coude foci used in radio-astronomy.

The Gaussian beam method assumes that the feed radiates a pattern which has no sidelobes so it is not adequate for a precise design, hence from now on we shall look at the radiation pattern in more detail. The general copolar pattern of a corrugated waveguide is shown in Fig. 5.5, indicating the E and H plane patterns. The difference between the two planes provides, of course, a measure of the amount of crosspolarisation present and this will be studied in the next section. Suffice it to say at this point that the crosspolar characteristics are determined mainly by the corrugation geometry, but this geometry has only a minor influence on the shape of the copolar patterns and so can be ignored for basic design purposes. The horizontal axes of Fig. 5.5 is plotted in normalised units $(D\sin\theta)/\lambda$, and so applies in general to any aperture diameter producing that level of crosspolarisation.

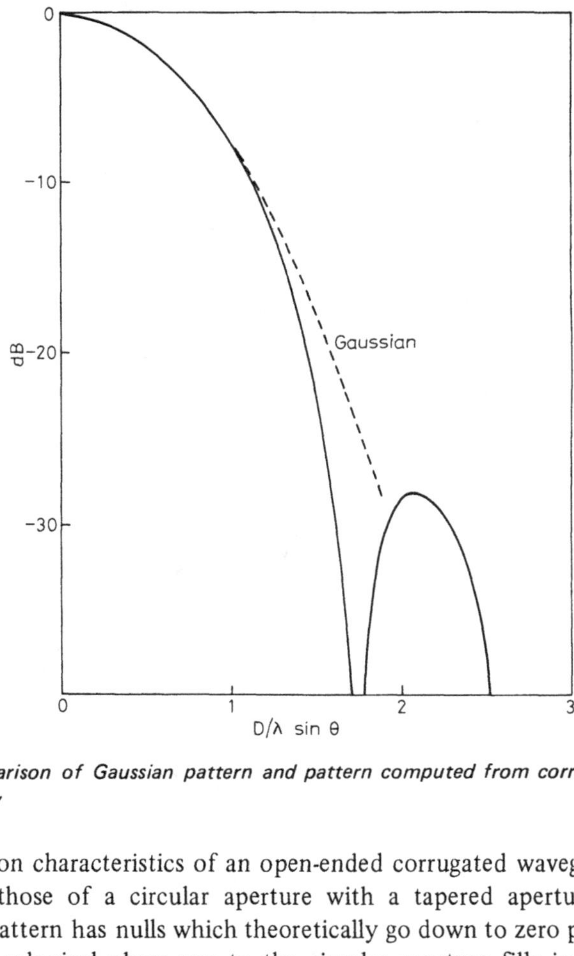

Fig. 5.4 *Comparison of Gaussian pattern and pattern computed from corrugated waveguide theory*

The radiation characteristics of an open-ended corrugated waveguide are similar in shape to those of a circular aperture with a tapered aperture distribution. The copolar pattern has nulls which theoretically go down to zero power level. The addition of a spherical phase cap to the circular aperture fills in the nulls. This occurs when the open-ended corrugated waveguide is transformed to a conical horn. Typical results are shown in Fig. 5.6. For even modest flare angles the pattern

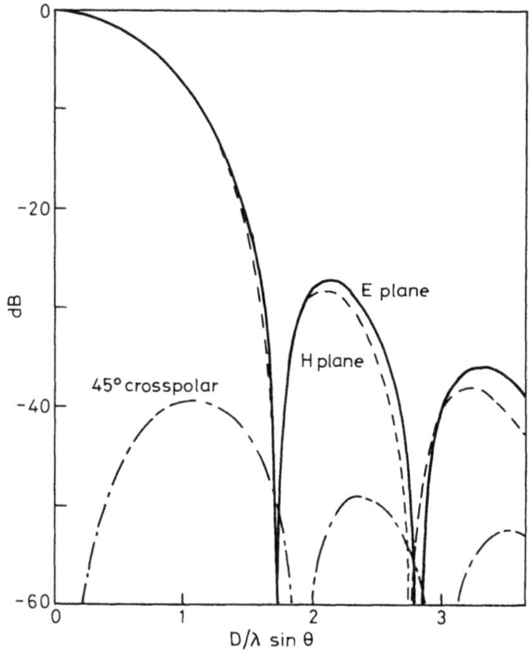

Fig. 5.5 *Radiation patterns of corrugated waveguide*

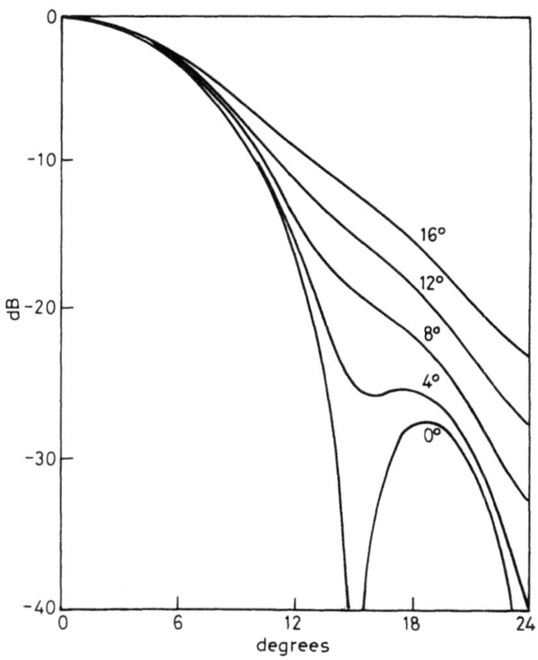

Fig. 5.6 *Influence of flare angle on copolar radiation pattern. $r_1 = 90$ mm, $s = 7{\cdot}5$ mm, 11 GHz, Parameter horn semi-flare angle*

differs substantially from the open-ended corrugated waveguide pattern. However the differences are again mainly outside the angular region used to illuminate a reflector. There is, therefore, considerable justification in modelling horns with semi-flare angles up to about 8° as open-ended waveguides, thus simplifying the analysis. It was shown by Thomas [149] that the patterns for narrow flare angle horns can be plotted in a normalised form using a spherical wave error Δ given by

$$\Delta = \frac{r_1}{\lambda} \tan \frac{\theta_0}{2} \tag{5.3}$$

The normalised patterns are shown in Fig. 5.7, taken from Thomas [149]. The aperture diameter is assumed to be greater than about two wavelengths so that the flange has little effect and for aperture diameters greater than about 4λ, the abscissa must be truncated at the value corresponding to $\theta = 90°$.

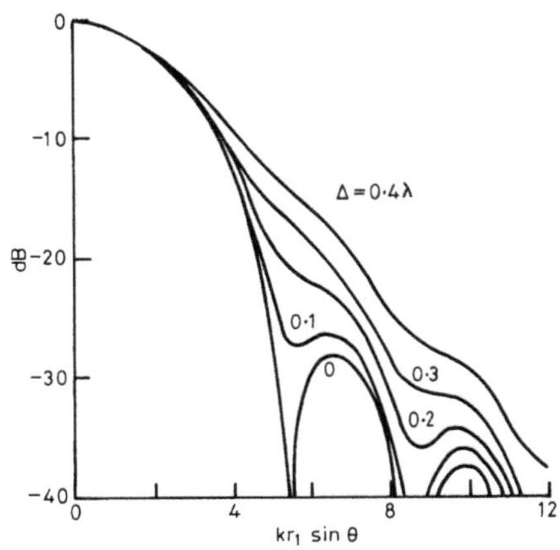

Fig. 5.7 *Normalised radiation patterns of small flare horns with Δ as parameter [149]*

Figure 5.7 applies for semi-flare angles up to about 20°. Above this angle the copolar radiation pattern becomes a function mainly of the flare angle and not of the aperture diameter. This is illustrated in Fig. 5.8, which shows the patterns for a 30° semi-flare angle horn of varying normalised slant lengths (assume frequency constant so balanced hybrid conditions maintained). For slant lengths of $R/\lambda > 4$ ($D/\lambda > 4$) the patterns have approximately similar beamwidths with two particular features. At the − 15 dB level the patterns cross at $\theta = 30°$, the semi-flare angle of the horn. This feature can be used as a rough design guide for wide flare angle horns. Secondly, the pattern around boresight loses its Gaussian-like shape. This is due to the spherical phase cap which causes the energy radiated from the centre of the aperture to be out of phase with the energy radiated from near the edge of the aperture.

The radiation patterns discussed above assume that the frequency is held constant. When the frequency changes, the normalised lengths change and the

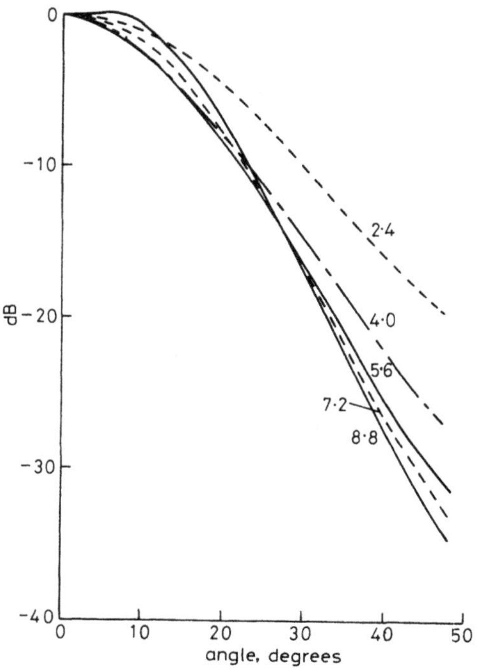

Fig. 5.8 *Copolar radiation patterns of 30° semi-flare angle corrugated horn. Parameter, normalised slant length R_0/λ*

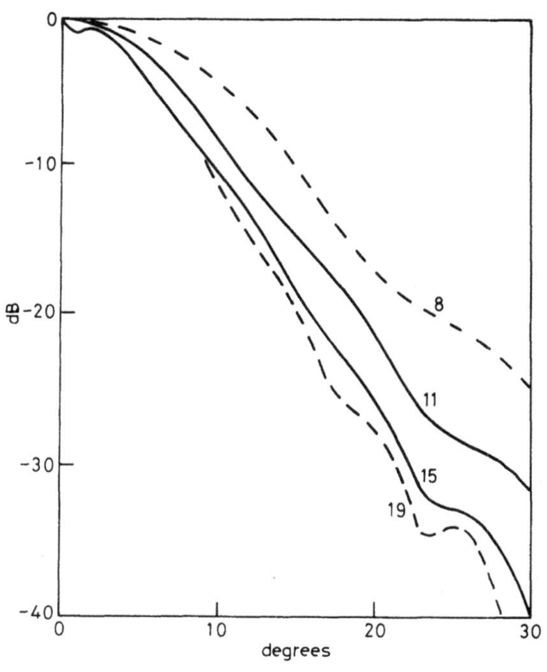

Fig. 5.9 *Influence of frequency on copolar radiation patterns of 12° semi-flare angle horn. $r_1 = 90$ mm, $s = 7{\cdot}5$ mm, Parameter: frequency (GHz)*

normalised slot depth changes. This moves the propagation characteristics off the balanced hybrid condition so that the pure hybrid mode conditions inside the horn no longer hold. However, experience shows that the corrugated surface is very tolerant of frequency changes, as far as the copolar patterns are concerned. Thus Figs. 5.5, 5.7 and 5.8 can be used to indicate the effect of frequency variations of about ± 20% for a practical horn. Over a wider frequency range, the copolar pattern must be computed including the corrugation geometry. Fig. 5.9 shows the patterns of a 12° semi-flare angle horn with a slot depth resonant at f_0 = 11 GHz. At 15 GHz (36% above f_0) the spherical phase error has introduced a small dip in the copolar pattern, while at 19 GHz (72% above f_0) the pattern is again smooth but the beamwidth is the same as at 15 GHz. The symmetry of the pattern has however been lost and the crosspolar level will be high.

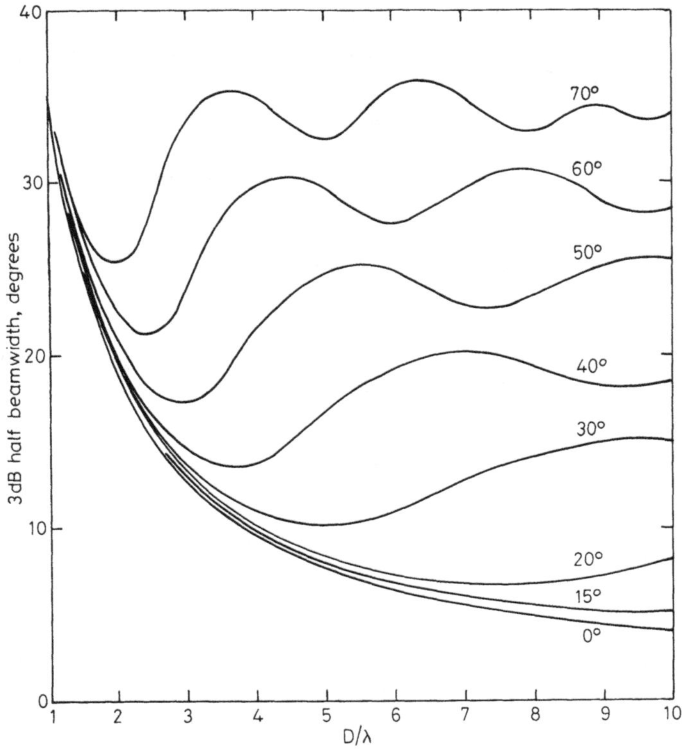

Fig. 5.10 – *3 dB half beamwidth against normalised aperture diameter. Parameter: semi-flare angle*

The designer of a feed is often given a copolar beamwidth specification and needs a horn which will meet the specification. Curves of beamwidth against normalised aperture size for fixed power levels, are useful to start the design process, and these are shown in Figs. 5.10, 5.11 and 5.12. These curves assume that the corrugation geometry has been adjusted to the balanced hybrid value at each point on the curve. Fig. 5.10, showing the – 3 dB half beamwidth illustrates many of the points discussed earlier in this section. The beamwidth of narrow flare angle horns and open-ended waveguides decreases monotonically with increasing aperture size, in contrast to the wider flare angle horns which oscillate about a fixed beamwidth.

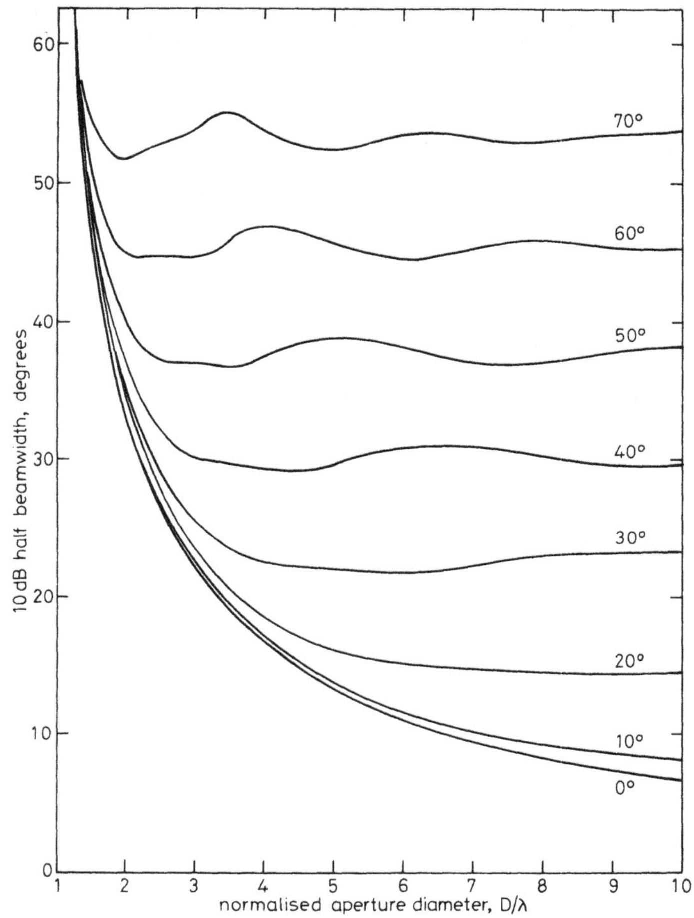

Fig. 5.11 *– 10 dB half beamwidth against normalised aperture diameter*

The oscillations are caused by the spherical phase factor. The same general behaviour is evident from Figs. 5.11 and 5.12, for the – 10 dB and – 20 dB half beamwidths, except that the oscillating behaviour is smoother. The wider angle horns have a peak gain when the beamwidth is a minimum. This occurs when the

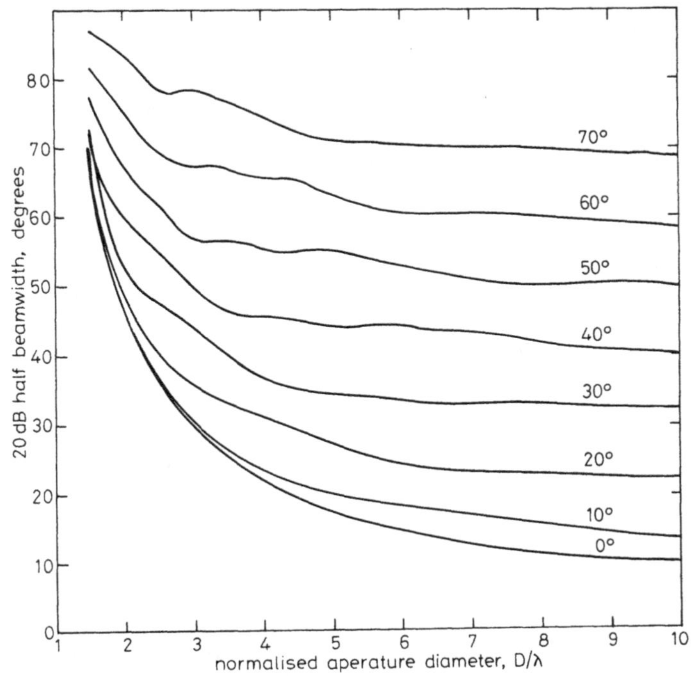

Fig. 5.12 *– 20 dB half beamwidth against normalised aperture diameter*

aperture is relatively small and the penalty for nearly constant beamwidth at larger aperture diameters is a poor aperture efficiency.

The copolar radiation patterns considered above have been the far-field patterns, computed using a space-harmonic model plus Fourier Transform pattern prediction. Sometimes it is necessary to know the radiation pattern at a distance nearer than the far-field, for instance if a subreflector is in the near-field of the corrugated feed. The near-field patterns can be computed using the spherical wave expansion technique (Chapter 4). However, with this technique the electric field model for the corrugations assumes a balanced hybrid mode and therefore cannot take account of the actual slot geometry. Near-field patterns for three horns with semi-flare angles 8°, 15°, 30° and $5{\cdot}4\lambda$ aperture diameter are shown in Figs. 5.13, 5.14 and 5.15, respectively. In the case of the 8° and 15° horns, the patterns are referred to the aperture, while with the 30° horn they are referred to the apex. The pattern shape is similar at all distances but the angle at which a given power level occurs changes substantially, particularly for the -20 dB power level. In terms of normalised quantities the classic far-field distance ($R = 2D^2/\lambda$) becomes $kR = 4\pi(D/\lambda)^2$. For the horn computed in Figs. 5.13–5.15, this gives $kR = 366$. It can be seen from the figures that even at this distance the radiated field is noticeably different from the $kR = \infty$ pattern. Note that the practical phase centre of the horn is not at the

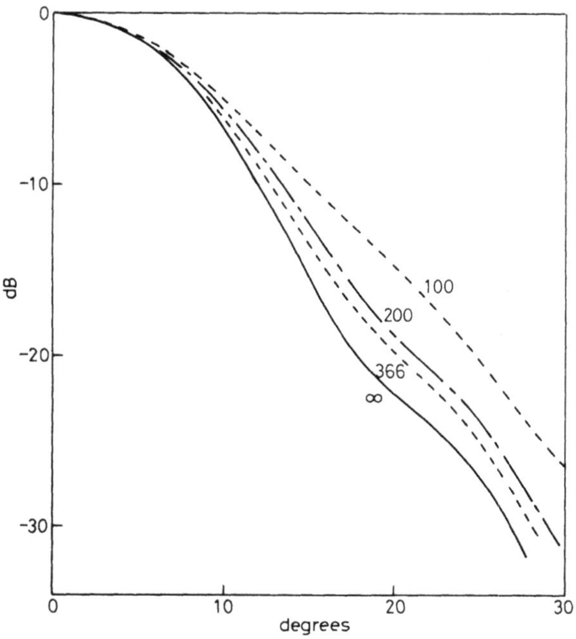

Fig. 5.13 *Near-field copolar radiation patterns for 8° semi-flare angle horn with normalised aperture diameter $D/\lambda = 5{\cdot}4$. Parameter: kR from aperture*

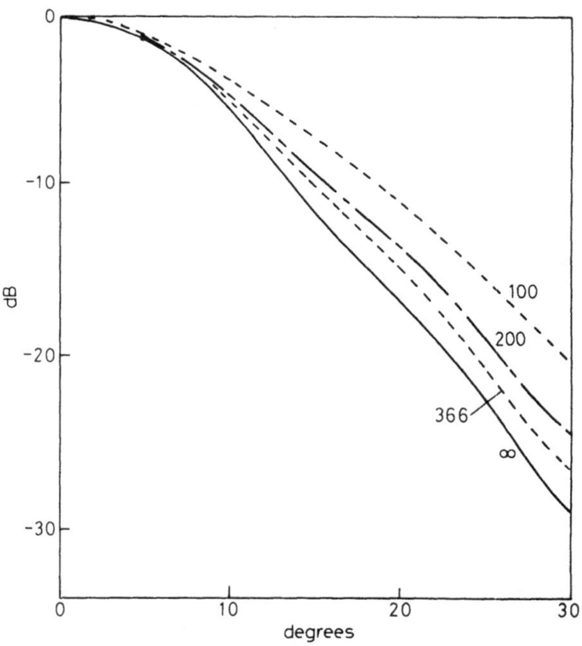

Fig. 5.14 *Near-field coplar radiation patterns for 15° semi-flare angle horn with $D/\lambda = 5{\cdot}4$. Parameter: kR from aperture*

aperture for the 8° and 15° semi-flare angle horns. This will modify the near-field patterns as a function of angle θ.

The presence of higher order modes may be wanted or unwanted. In the former case, where they are used for tracking or as part of a 'matched feed' [132], the modes under consideration are the H_{01}, E_{02} and HE_{21}. All these mode have copolar patterns with a null on boresight and change phase through the boresight.

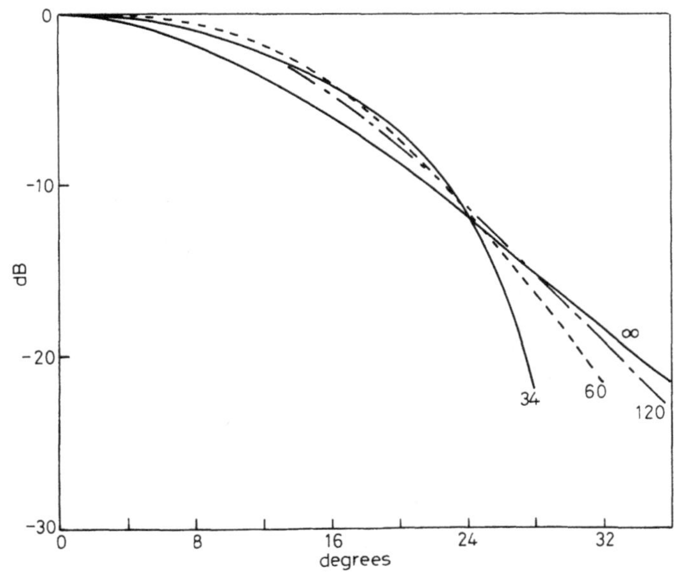

Fig. 5.15 *Near-field copolar radiation patterns for 30° semi-flare angle horn with D/λ = 5·4. Parameter: kR from apex*

Since the tracking capability is usually an 'extra', the precise pattern shape will not be a fundamental specification for the designer. Typical patterns are shown in Fig. 5.16. The three patterns have been normalised to the same peak level, and it can be seen that they have the same shape in the angular region of interest.

When the higher order modes are present but unwanted, the copolar pattern can be distorted. The two modes of interest are the HE_{12} and EH_{12} modes. Both these can be excited in the throat and flare regions of the horn due to changes in the horn cross-section with axial distance (see Chapter 4). Typical copolar patterns of the two modes are shown in Fig. 5.17. The HE_{12} modes has a maximum on boresight, good pattern symmetry and a first sidelobe nearly as strong as the main lobe. It occurs at an angle corresponding to the HE_{11} pattern level typically used to illuminate the edge of a reflector (see Fig. 5.5). The EH_{12} mode pattern has a null on boresight, a peak at about the same angle as the HE_{12} mode and a difference between the *E* and *H* plane pattern which is sufficient to give rise to a significant level of cross-polar radiation. The effect of adding an HE_{12} mode pattern to the

HE_{11} mode pattern is shown in Fig. 5.18. These are E plane patterns, the H plane patterns are influenced in the same way. The amount of distortion depends not only on the level of HE_{12} power but also on the phase angle between the two modes. The sidelobe level may be raised or lowered. The presence of HE_{12} mode

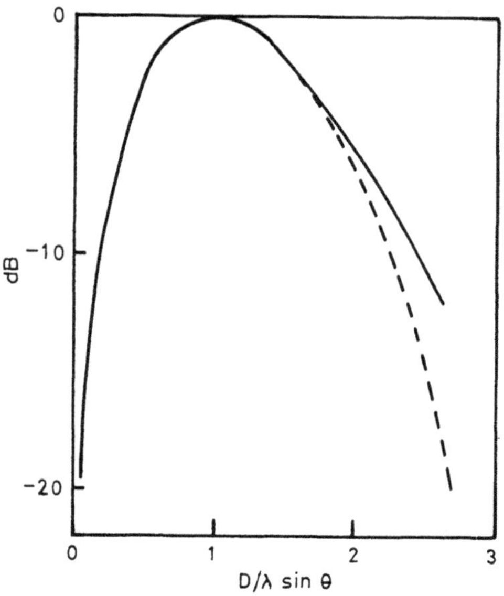

Fig. 5.16 *Copolar radiation patterns of* ——— *HE_{21} and E_{02} modes,* – – – *H_{01} mode*

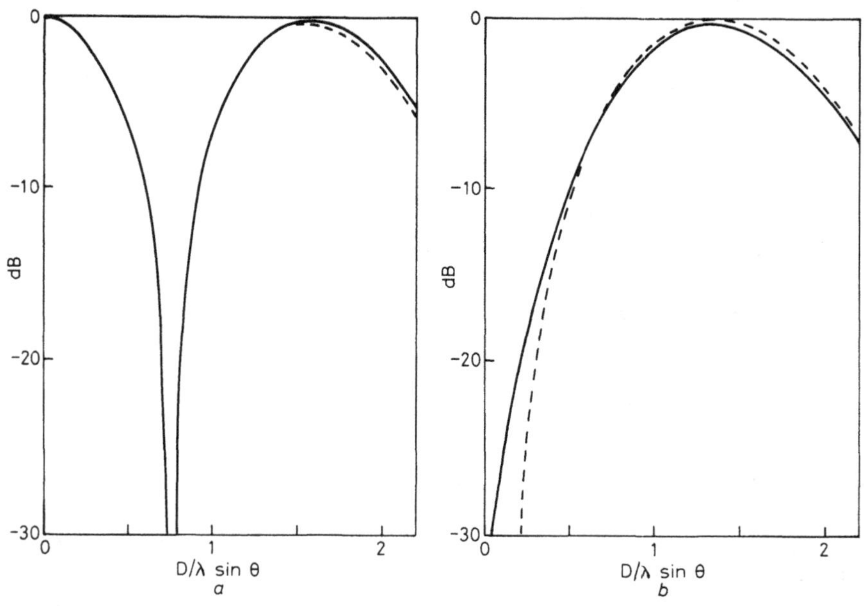

Fig. 5.17 *Copolar radiation patterns for, a HE_{12} mode and b EH_{12} mode*
——— E plane – – – H plane

distortion on a measured pattern can be confirmed by measuring over a frequency band in order to detect the phase changes. The crosspolar patterns are virtually unaffected by the presence of the HE_{12} mode power. The EH_{12} mode, at the same relative level as the HE_{12} mode, influences the copolar patterns in a similar way, except that there is a 180° phase difference between the EH_{12} E and H plane

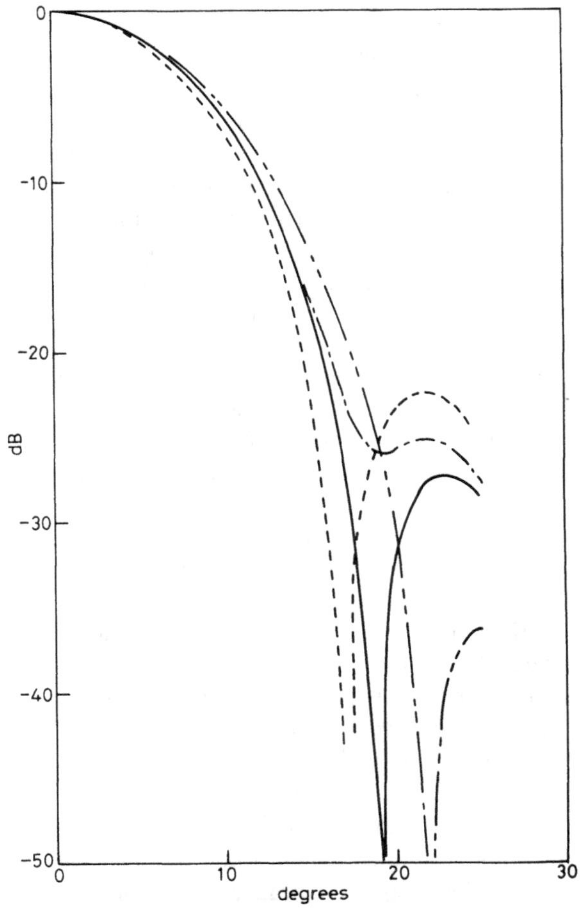

Fig. 5.18 *Addition of − 25·4 dB of HE_{12} mode to HE_{11} mode of an open-ended corrugated waveguide with r_1 = 81 mm, s = 8 mm, at 10 GHz*
——— HE_{11} mode only – – – ∠0° HE_{12} mode – · – ∠90° HE_{12} mode – · · – ∠180° HE_{12} mode

patterns. This means that the two sum patterns are not affected in the same sense, i.e. for zero degree phase angle, the $HE_{11} + EH_{12}$ E plane corresponds to the $HE_{11} + HE_{12}$ (zero degree phase) E plane pattern, but the $HE_{11} + EH_{12}$ H plane pattern corresponds to the $HE_{11} + HE_{12}$ (180 degree phase) E plane pattern. In addition the crosspolar level will be significantly raised. Thus examination of distorted measured patterns can lead to knowledge of whether either HE_{12} or EH_2 modes are present.

5.3 Crosspolar radiation characteristics

We shall now discuss the design of corrugated horns to produce low crosspolarisation as defined in the last section. We shall assume that the horn is propagating an HE_{11} mode and that the peak crosspolar power occurs in the 45° plane. The crosspolar power can be attributed to four sources: the intrinsic crosspolarisation of the corrugated structure; the higher order modes generated by the throat region; the higher order mode conversion along the length of the horn and direct radiation from the flange. The second and third factors will be considered in later sections, the fourth factor is only of concern for small horns (aperture diameter less than two wavelengths), while the first factor is the subject of this section. In general it can be stated that, for most corrugated horns, it is the intrinsic crosspolarisation which will dominate the total radiated crosspolarisation. Only when the intrinsic level has been reduced to a sufficiently low level, will the other factors become significant.

The level of crosspolar power radiated at any frequency is determined by the size of the aperture and by the geometry of the corrugations, particularly the slot depth. Since the aperture size is the factor governing the copolar characteristics, the crosspolar designer normally has no control over the aperture size. He can, however, adjust the corrugation geometry to optimise the crosspolar radiation over the design band of frequencies. In order to study the influence of the corrugations on the radiation characteristics we need a theoretical model which includes the exact geometry of the corrugations. For this purpose we will use the space harmonic model (see Appendix) with one pair of space harmonics in addition to the fundamental HE_{11} mode in the inner region and one pair of slot modes. Experience has shown that one pair of space harmonics is adequate for design and study purposes. Including further space harmonics makes only a few percent difference to the power level, under the worst circumstances.

The space harmonic model allows for an accurate description of the electric and magnetic field throughout the waveguide, in contrast with the equivalent surface impedance model. However accuracy is traded for simplicity. Before considering the exact model let us review the main features of the crosspolar radiation from a corrugated horn or waveguide, radiating the HE_{11} mode. The crosspolarisation is zero on boresight and also zero along the principal planes. It rises to a peak along the 45° planes, Fig. 5.19 (note that the crosspolarisation level is always referenced to the copolar boresight). The first crosspolar lobe peaks at an angle corresponding to a copolar power level of about -9 dB. A result which is valid for most narrow flare angle corrugated horns. For wide flare angle horns, the peak level moves inwards towards the boresight direction. The significance as far as the user is concerned is that the peak crosspolar level lies within that portion of the main copolar beam which illuminates either the main reflector or the sub-reflector. Hence the crosspolarisation is transformed through the reflector system to appear in the far-field of the reflector antenna. For this reason the peak crosspolar level is a significant design parameter. The variation of the peak level with normalised frequency (f/f_0) is shown in Fig. 5.20. An equation for the peak crosspolar level is given in Section 3.5 (eqn. 3.82), but because of the model, it does not take account of the geometry of the corrugations. However, it does indicate that the bandwidth over which the peak crosspolar power lies below a given value is proportional to the aperture diameter of the horn.

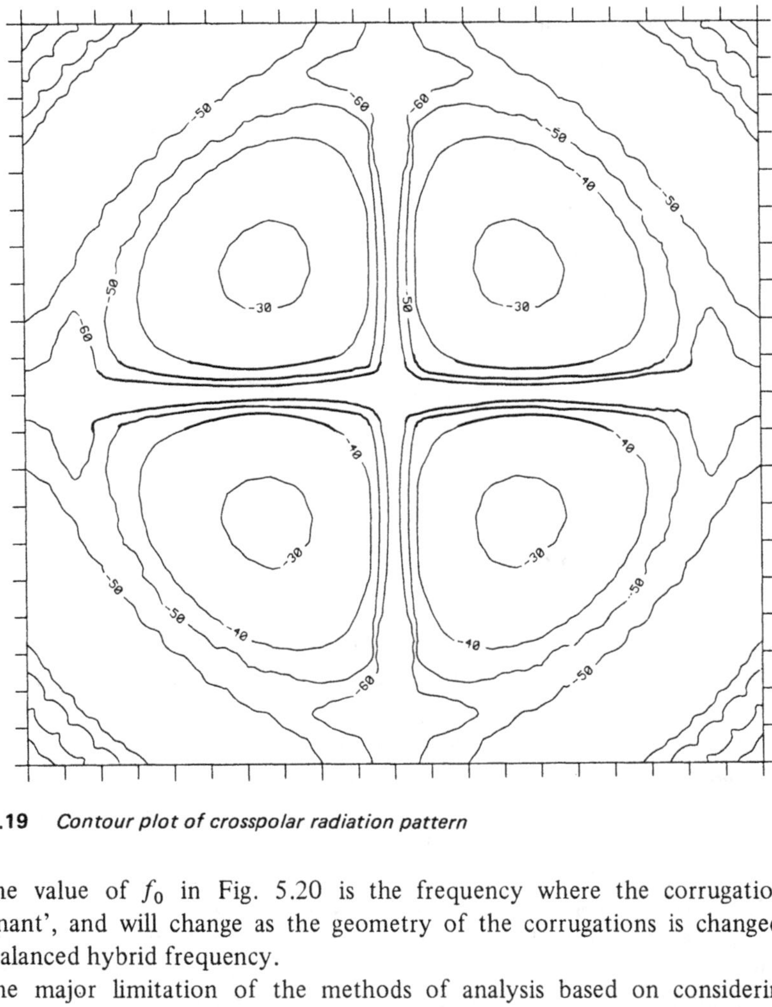

Fig. 5.19 *Contour plot of crosspolar radiation pattern*

The value of f_0 in Fig. 5.20 is the frequency where the corrugations are 'resonant', and will change as the geometry of the corrugations is changed. It is the balanced hybrid frequency.

The major limitation of the methods of analysis based on considering the corrugations as an admittance surface is in the precise determination of f_0. The simple model assumes that the normalised susceptance of the corrugations can be represented by

$$Y = \cot(ks) \tag{5.4}$$

and that the hybrid factor is given by (Section 3.2, Table 3.2),

$$\bar{\Lambda} = 1 - x_{0m}^2 \left[\frac{Y}{2} \frac{1}{kr_1} - \frac{Y^2}{8}(4 + x_{0m}^2)\left(\frac{1}{kr_1}\right)^2 \cdots \right] \tag{5.5}$$

Further the crosspolar radiated field is proportional to the factor:

$$\frac{\bar{\Lambda} - 1}{\bar{\Lambda} + 1} \tag{5.6}$$

From the above equations it would appear the crosspolarisation will be zero when the slot depth, s, is $\lambda/4$ and the hybrid factor is unity. This is not borne out in practice, except for large diameter horns. A better model is obtained by using the full surface impedance approach of eqn. 3.26 but even here the longitudinal geometry of the corrugations are not treated in the model.

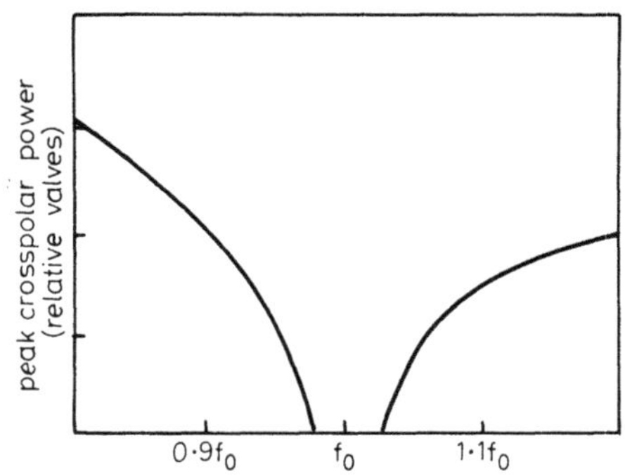

Fig. 5.20 *Sketch of peak crosspolar radiation against normalised frequency*

The above discussion has highlighted the problem of using a simple theoretical model to predict the crosspolarisation characteristics. It means that the hybrid factor, $\bar{\Lambda}$, which is so useful in explaining qualitatively the propagation and radiation behaviour of corrugated waveguides cannot be used for precise crosspolar design calculations. Computations with the space-harmonic model show that the value of $\bar{\Lambda}$ to produce minimum crosspolarisation is always greater than unity, but there is as yet, no simple analytic relationship to enable the value to be predicted from the corrugation geometry.

It is the presence of space harmonics in the corrugated horn or waveguide which modifies the balanced hybrid frequency, f_0. This is shown in Fig. 5.21, where the power in the HE_{11} mode of a relatively small diameter waveguide has been divided into the components from the individual $-1, 0, +1$ space-harmonics. The frequency for minimum crosspolarisation depends not only on the fundamental harmonic, $N = 0$ (which is the only one considered in the surface impedance model) but on the higher order harmonics. The power in the space harmonics becomes less significant as the diameter of the horn is increased. However, they retain their influence on the balanced hybrid frequency long after they have ceased to be significant for copolar predictions. This is because the fields of the space-harmonics are confined near to the corrugated surface and this surface controls the filter-like properties of the crosspolar frequency characteristic.

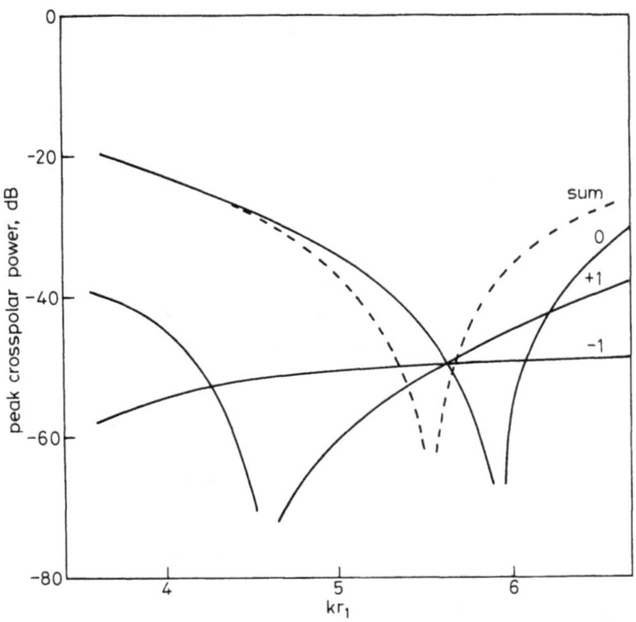

Fig. 5.21 *Influence of space harmonics on peak crosspolar radiation of a corrugated horn with* $r_1/r_0 = 0{\cdot}8$

We shall now study how the corrugation geometry effect the balanced hybrid frequency. Consideration of the space-harmonic formulation (see Appendix) shows that the following parameters are significant:

$$\frac{D}{\lambda}, \frac{s}{\lambda}, \frac{b}{\lambda}, \frac{t}{b}$$

that is, the normalised diameter (D/λ), normalised slot depth (s/λ), normalised slot width (b/λ) and the ratio of ridge width to slot width (t/b). A version of our space-harmonic program was written to undertake a parametric study of the normalised quantities for radiation from an open-ended corrugated waveguide. The pertinent results are plotted in Figs. 5.22, 5.23, 5.24 and 5.25. They show the normalised slot depth which gives minimum crosspolar power against the normalised aperture diameter for either fixed b/λ or fixed t/b. The slot widths are varied between $0{\cdot}1\lambda$ and $0{\cdot}3\lambda$ and the ridge width to slot width ratios between 0·1 (narrow ridge) to 1·0 (equal slot/ridge width). This spans the range between 2 slots per wavelength and 9 slots per wavelength.

It should be emphasised that throughout this range the form of the peak cross-polarisation is as shown in Fig. 5.20. One of the most common questions asked about corrugated feed design is 'how many slots per wavelength?'. The answer from the point of view of the radiation characteristics is that, within a wide range it does not matter as long as the slot depth is adjusted according to Figs. 5.22–5.25. A number of features are evident from the figures. The slot depth is always deeper for

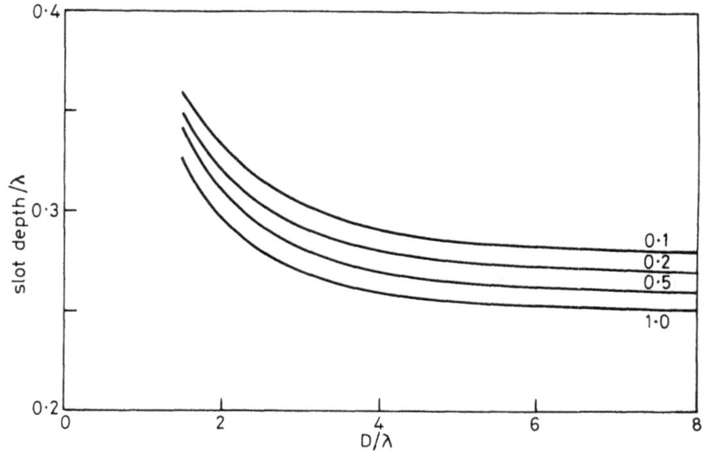

Fig. 5.22 *Slot depth for minimum crosspolarisation against normalised aperture diameter. Slot width = 0·1λ. Parameter: ridge width to slot width ratio*

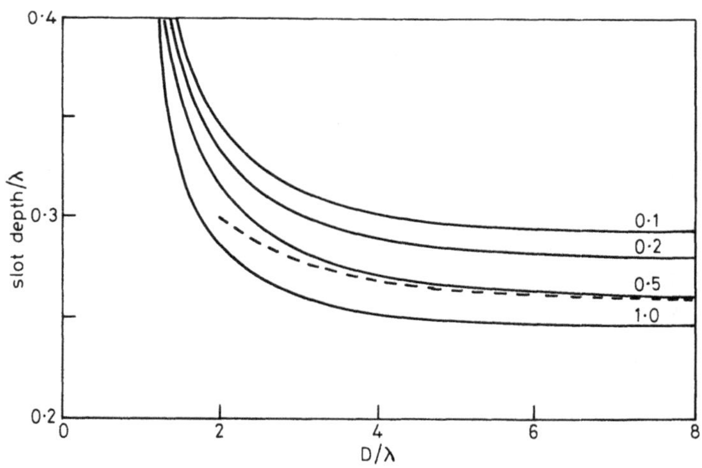

Fig. 5.23 *Slot depth for minimum crosspolarisation. Slot width = 0·2λ. Parameter: ridge width to slot width ratio. — — — surface impedance result*

small aperture diameters than for large aperture diameters. Only for large diameters do the slot fields follow trigonometric function relations, then the slots can be considered to belong to a plane surface. For aperture diameters under 2λ the slot depth to produce balanced-hybrid conditions is much deeper than the nominal quarter wavelength deep slot.

The second observation which can be deduced from the figures is the influence of the longitudinal geometry of the corrugations in determining the balanced-hybrid frequency, even for large aperture sizes. Also it is clear that the ratio of the ridge width to the slot width is more important than the normalised slot width. Small values of t/b require deeper slots than when the slot and ridge have equal width.

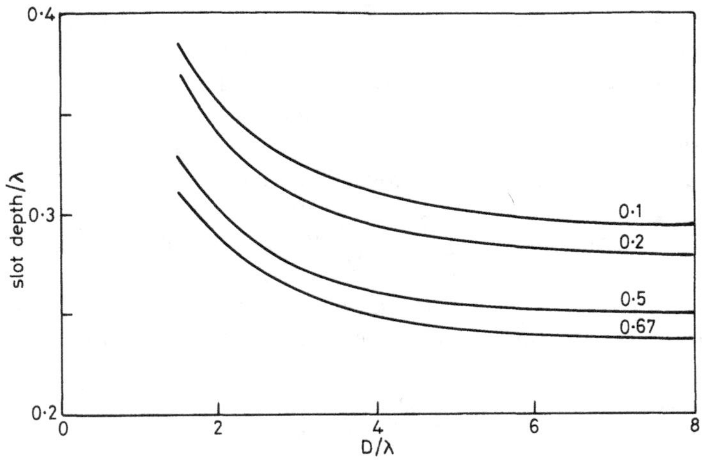

Fig. 5.24 *Slot depth for minimum crosspolarisation. Slot width = 0·3λ. Parameter ridge width to slot width ratio.*

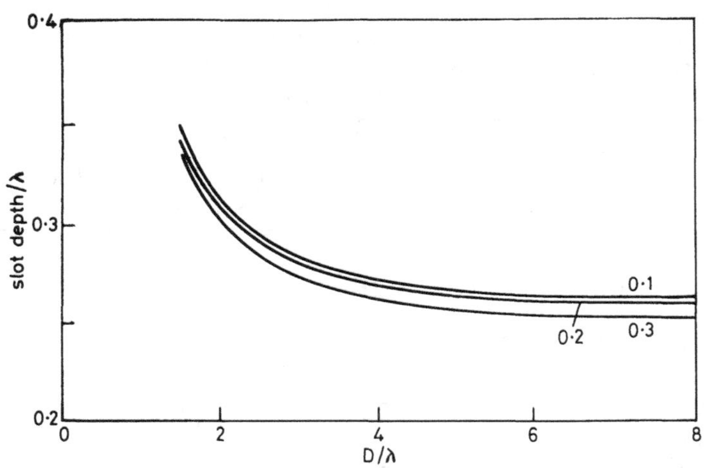

Fig. 5.25 *Slot depth for minimum crosspolarisation. Ridge width to slot width ratio = 0·5. Parameter: slot width, λ*

Slots which are relatively wide, e.g. 0·3λ, Fig. 5.24, show more difference between the minimum and maximum ridge/slot width ratios. The maximum value which could be computed was $t/b = 0{\cdot}67$. For larger values numerical instability in the computations occurred. Note that for this value the normalised slot depth is less than $\lambda/4$. The fields around the corrugations are now so different from the values assumed in a surface impedance model that concepts such as inductive and capacitive slots have little meaning. In passing we note that wide slots are important at millimetric and sub-millimetric frequencies where the practical problems of making corrugated horns demand that there be few corrugations per wavelength.

Fig. 5.23 also shows the result produced by the surface impedance model of eqns. 3.16–3.26. This is similar to the $t/b = 0{\cdot}5$ curve. Its significance lies in the fact that most corrugated feeds which have been designed so far would lie on this curve (i.e. feeds with $D/\lambda > 3$, $b/\lambda < 0{\cdot}2$ and $t/b \approx 0{\cdot}5$). Thus a surface impedance model, in which the slot fields are represented by Bessel functions, does in fact give a good estimate of the frequency at which minimum crosspolarisation occurs, as long as these restrictions are noted. Indeed it is possible to use the even simpler model of eqn. 5.4 by computing the crosspolarisation assuming $\lambda/4$ slots, but then using Figs. 5.22–5.25 to obtain the real slot depth.

Experimental verfication of Figs. 5.22–5.25 has been obtained by comparing the predicted crosspolar minimum with values measured on a number of feeds with a wide range of geometries. The comparison is shown in Table 5.3. The generally good agreement between theory and experiment is adequate confirmation of the parametric study.

The discussion above specifically referred to narrow flare-angle horns or open-ended waveguides. However, as Table 5.3 shows, the values apply to wider angle horns as long as the electric depth of the slots at the aperture is used. This value is not always easy to predict. For horns with semi-flare angles up to about 15°, the

Table 5.3 *Comparison of theoretical and experimental results for the frequency of minimum crosspolar radiation: Measured and computed results are for the frequency where minimum peak cross-polar power occurs*

Horn	$\frac{D}{\lambda}$	slot width λ	$\frac{\text{ridge w}}{\text{slot w}}$	Theory ($K = L = 1$) (slot depth)/λ	Measurement (slot depth)/λ
Open-ended waveguide	1·64	0·33	0·16	0·36	0·35
Profiled horn (9.7 GHz)	2·88	0·1	0·33	0·29	0·28
12° linear horn (8.7 GHz)	3·68	0·23	0·19	0·29	0·28
20° linear horn (11° GHz)	6·88	0·18	0·3	0·27	0·26
30° linear horn (9.8 GHz)	5·0	0·26	0·31	0·28	0·26
5·3° linear horn (14.1 GHz)	1·36	0·12	0·21	0·39	0·39

electrical depth can be taken as the depth in the centre of the slot, see Fig. 5.26; for larger flare angles a mean depth is best chosen and must partly be done by trial and error. If the slots are cut perpendicular to the axes, the central slot depth can be

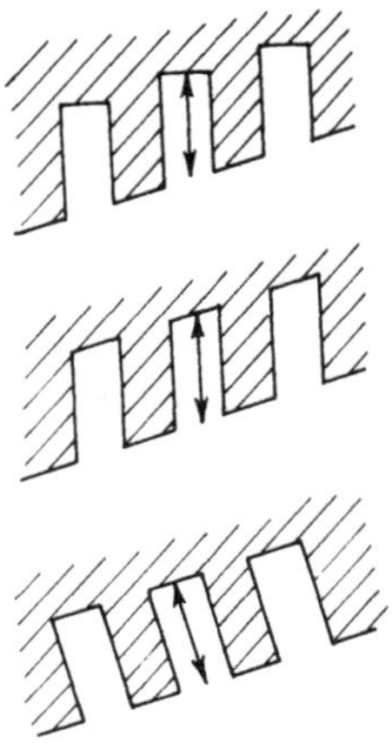

Fig. 5.26 *Orientations of slots*

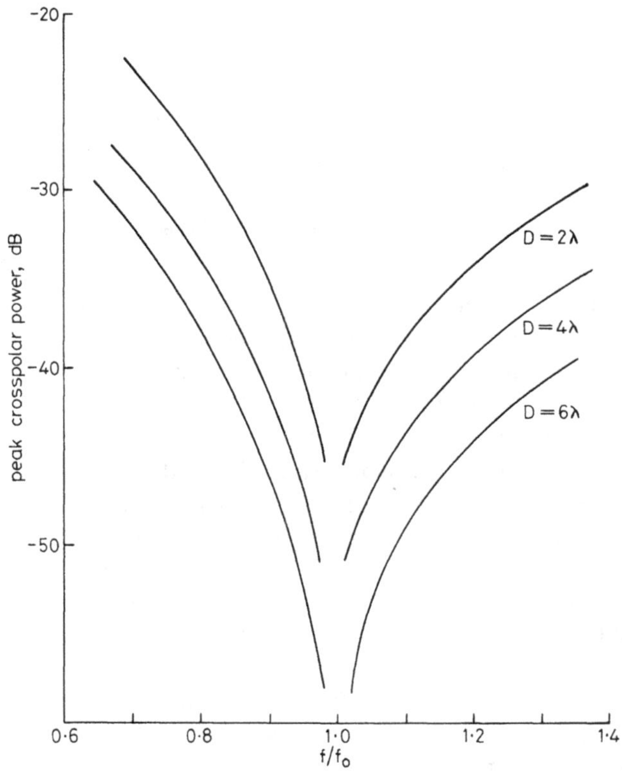

Fig. 5.27 *Peak crosspolar radiated power as a function of f/f_0 where f_0 is the frequency for minimum crosspolarisation. Parameter: normalised aperture diameter*

used, if they are cut perpendicular to the horn flare angle a value about 20% less than the slot depth gives a reasonable estimate. The value is easier to estimate if the slot width is narrow. However, experience shows that corrugated horns work just as well whichever style of slot is chosen, so the choice can be made on mechanical grounds rather than electrical grounds.

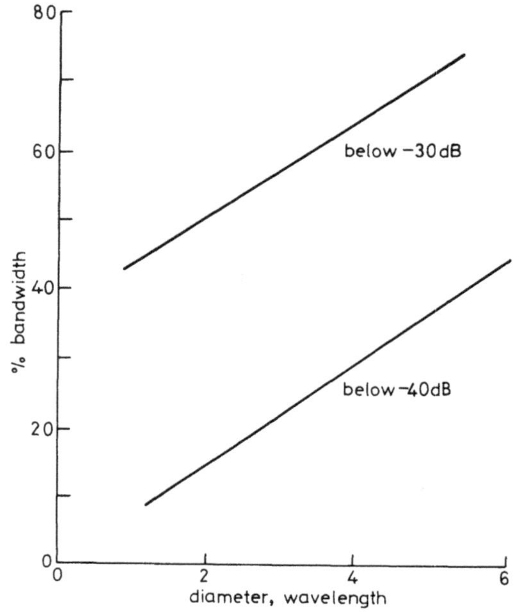

Fig. 5.28 *Theoretical percentage bandwidth for peak crosspolar level below − 30 dB and − 40 dB*

When the balanced hybrid frequency, f_0, has been determined or chosen for a particular feed, the crosspolar bandwidth can be computed. Theoretical curves showing the peak crosspolar power for three normalised aperture diameters are shown in Fig. 5.27. From these curves, Fig. 5.28 has been obtained, showing the theoretical percentage bandwidth below which the peak crosspolar level is less than − 30 dB or less than − 40 dB. We should stress that both curves are the theoretical levels assuming all the crosspolarisation is generated by the intrinsic geometry of the waveguide. If crosspolarisation is present due to higher order modes or due to the flange, it may well dominate the intrinsic crosspolarisation. The crosspolar bandwidth will sometimes be the factor which determines the usable bandwidth of the feed because it can be much less than the copolar bandwidth. Attempts have been made to increase the crosspolar bandwidth of conventional corrugated horns but without success. However the dual-depth horn exhibits two bands of frequency in which the crosspolarisation is low, see Section 5.8.7.

Before finishing this section, a brief reference will be made to the near-field crosspolarisation patterns. It was shown in Section 5.2 that the copolar patterns do not change their shape until the observation point is very near to the horn. The crosspolar results show a similar behaviour in that the peak crosspolar level remains more or less constant from the aperture to the far-field. Sample results for a 12°

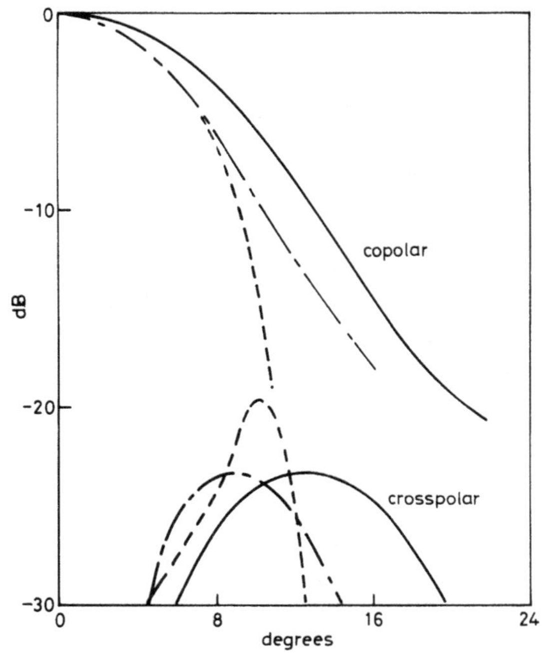

Fig. 5.29 *Near-field radiation patterns for 12 degree semi-flare angle horn with $kR_0 = 81$.*
— — — $kR = \infty$ — · · — $kR = 280$ from apex. — — — $kR = 81$ from apex

semi-flare angle horn are shown in Fig. 5.29. The angle at which the peak crosspolar level occurs moves inwards as the distance from the horn to the far-field is reduced. Thus the horns can be operated well into their near-field region without significantly altering the crosspolar radiation patterns.

5.4 Flare section of horn

We have discussed in the previous sections the design of the aperture of the horn to achieve a specified copolar radiation pattern and the design of the corrugation geometry at the aperture to achieve a specified crosspolar characteristic. This section will deal with the design of the flared section between the throat and the aperture. In some cases the copolar pattern specifications will determine the flare angle, see Section 5.2, but there are three other considerations:

(i) the mechanical constraints on the length;
(ii) the position and movement of the phase-centre;
(iii) the mode conversion which takes place along the flared section.

For a given aperture diameter and a given throat design the safe solution is to have a long horn with a narrow flare angle. This approach has been adopted in many large dual-reflector earth-station antennas. There are however many cases where the long length and consequent weight are not practical, and in these circumstances it is important to know the electrical consquences of changes from the ideal.

Phase centre

The phase centre of any horn is the point on the axis of the horn which is the centre of curvature of the phase front. It is usually found by measurement or by deduction from the computed radiation pattern. In either case the assumption is made that the fields can be considered to be spherical waves and hence, given the amplitude and phase patterns, the centre of the sphere (the phase centre) can be calculated. It is possible to find the practical phase centre by placing the test horn on a turntable and recording the phase pattern for various rotation points along the axis of the horn. When the phase patterns are constant, with changing azimuth angle, the test horn is rotating about its apparent phase centre.

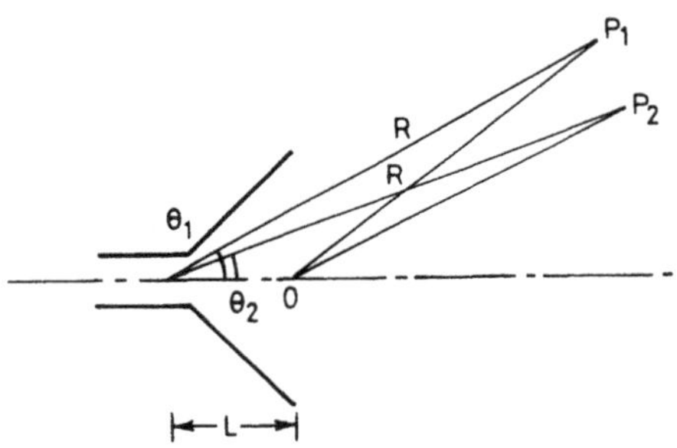

Fig. 5.30 *Geometry for calculation of phase centre*

The apparent, theoretical, phase centre can be computed from the amplitude and phase of the radiated field at any distance from the horn by assuming that the radiation at two adjacent points, in the same plane, emanates from a common point on the axis of the horn. Let the two points, P_1 and P_2, be at angles θ_1 and θ_2 and distance R from the apex of the horn, Fig. 5.30, and let the apparent phase centre be at 0, distance L from the apex. If the difference between the radiation phase angles at P_1 and P_2 is $\delta\psi$ then, by geometry, the distance L is given by

$$L = \frac{R\left\{-\tfrac{1}{2}(\cos\theta_2 + \cos\theta_1) \pm \left[\cos\theta_2\ \cos\theta_1 - 1 + \dfrac{X^2}{4} + (\cos\theta_2 - \cos\theta_1)^2/X^2\right]^{1/2}\right\}}{[(\cos\theta_2 - \cos\theta_1)^2/X^2 - 1]} \tag{5.7}$$

where $X = \delta\psi/kR$.

The positive sign is for $\delta\psi$ positive and the negative sign for $\delta\psi$ negative. In the far-field, $R \to \infty$ and the above equation simplifies to:

$$\frac{L}{\lambda} = \frac{\delta\psi}{\cos\theta_2 - \cos\theta_1} \tag{5.8}$$

These equations enable the apparent phase centre to be computed at each point in the radiated field.

This assumes that the phase centre is well defined, i.e. that the fields at any point in space are spherical waves. Unfortunately this is not true. There is no single phase centre in most horns. The phase centre varies with the plane of polarisation, the angle of boresight, the frequency and the distance from the horn. There is one important exception; if the horn is an open-ended waveguide the phase centre remains essentially fixed at the aperture of the waveguide. This fact has partly given rise to the profiled corrugated horn as a means of achieving the ideal situation.

To return to the phase centre in linear flare angle horns, it is difficult to lay down quantitative guidelines for the behaviour of the phase centre, so what follows is a qualitative discussion of the affects. Each horn needs to be treated individually and the phase centre computed for the desired radiation conditions. The phase centre in the E and H planes, will, in general, be different. However, for low cross-polar designs the E and H plane patterns are nearly coincident and hence the phase centres can be assumed to be coincident.

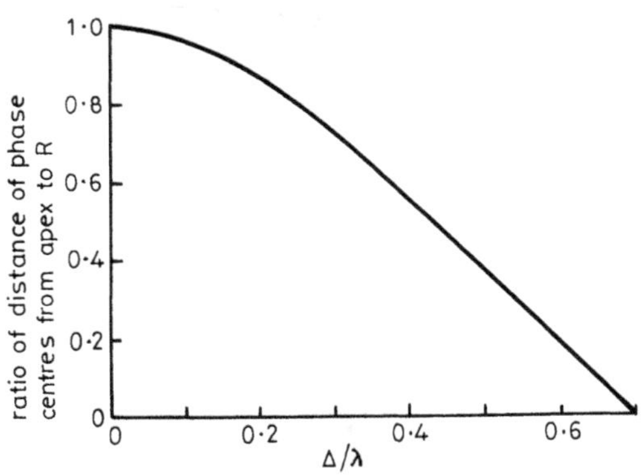

Fig. 5.31 *Normalised distance of horn phase centre from apex against Δ/λ [149]*

The horn flare angle and the frequency of operation also determine the position of the phase centre. For wide flare angle horns the phase centre is near the apex, so for horns with intermediate angles between a waveguide and a wide flare angle horn the phase centre is in between the apex and the aperture. The far-field position can be conveniently expressed in terms of the sperical phase factor Δ, eqn. 5.3. Fig. 5.31 shows the position for narrow flare angle horns. This gives the position for an observer in the far-field and along the boresight of the horn. The phase centre varies with angle of boresight because the radiated field is a summation of spherical modes. The variation with azimuth angle is plotted in Fig. 5.32 for horns with semi-flare angles of 8° and 15° and 5·4λ aperture diameter.

The distance of the observation point from the horn also determines the phase centre. The position in the near-field is not the same as the far-field position. At the aperture the horn fields are nearly spherical so the phase centre at the aperture is always at the apex of the horn. As the distance from the horn is increased the phase

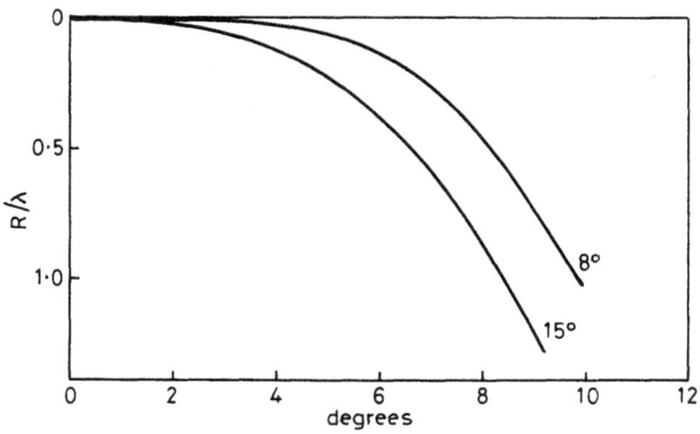

Fig. 5.32 *Relative phase centre movement with angle for horn with normalised aperture diameter = 5·4λ. Parameter: semi-flare angle*

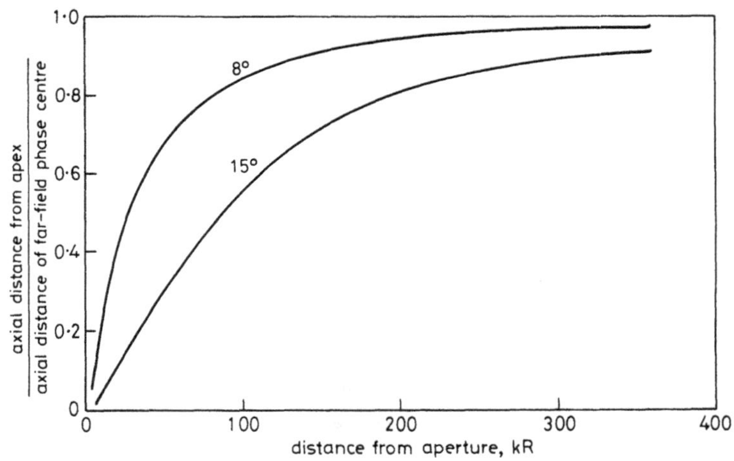

Fig. 5.33 *Position of near-field phase centre relative to far-field phase centre. Parameters as Fig. 5.32*

centre will move outwards from the apex until it reaches the far-field value given by Fig. 5.31. The position is shown in Fig. 5.33 for the 8° and 15° semi-flare angle horns of Fig. 5.32. Again the 8° semi-flare angle horn reaches the far-field value sooner than the 15° semi-fire angle horn. This fact is particularly relevant when the horn is in the near-field of a reflector and means that the horn and reflector need to be closer than is apparent by considering the far-field patterns.

Mode conversion along the horn

The mode conversion along the horn needs to be kept to a very low level. The theory of mode conversion, as applied to wide flare horns, has been studied in Chapter 4 using spherical modes. Using cylindrical modes, Dragone [56] showed that the mode conversion in small flare angle, large diameter horns with constant surface reactance, can be expressed in the form:

$$M^2 = \frac{1{\cdot}357 \times 10^{-2} \tan^2\theta}{\tan^2(ks)} \tag{5.9}$$

where M is the mode conversion coefficient. This implies that it is necessary to keep the flare angle small or to maintain a high wall reactance such that $\tan^2 ks \rightarrow \infty$. The spherical mode model shows that the mode conversion is a function of $\partial\theta/\partial z$, i.e. the rate is zero in a constant flare angle horn and so indicates that the lowest mode conversion is obtained with narrow, linear flare angle, constant depth slots. The latter criteria conflicts with the need to transfer power from a smooth-wall waveguide, which means that slots near the throat must have varying slot depths (see next section). The former criteria tends to suggest that only linear flare angle horns should be used. However, there are some cases where there are overriding reasons for not using a linear flare angle horn.

Profiled horn

As has already been discussed, the phase centre of the horn can only be fixed with changing frequency by radiating from an open-ended waveguide. A profiled horn enables this to happen. Another very useful reason for choosing profiled horns is to reduce the physical length of the horn and produce a compact horn. This is particularly suitable for spacecraft applications. The form of the profile is sketched in Fig. 5.34, and is a sine-squared type of profile. The concept is similar to the profiled waveguide tapers developed for the TE_{01} overmoded waveguide. The total length of the horn can be reduced by up to one third of the length required for a linear taper, the limit of reduction being determined by the acceptable mode conversion level. A short section of constant diameter corrugated waveguide is added to the aperture to ensure that the phase centre conditions are met. A suitable form for the profile is given by:

$$r(z) = r_{11} + (r_{12} - r_{11})\left[\frac{z}{L}(1 - A) + A\sin^2\left(\frac{\pi}{2}\frac{z}{L}\right)\right] \tag{5.10}$$

The parameter A determines the amount of profiling. When $A = 0$ a linear taper results, when $A = 1$ a sine squared profile results. A value of $A = 0.7$ gives a reasonably gradual change in diameter. These types of horn have been built and tested and found to work well, see Section 5.8.7.

The mode conversion along the horn due to changing flare angle is analysed in Chapter 4. It is shown that low mode conversion to the HE_{12} mode occurs when the profile has a maximum flare angle nearer to the aperture than half way along the horn. This means that the profile is non-symmetrical and for the first half of the horn L_1, Fig. 5.34, the inner radius increases slowly but in the second half of the horn, L_2, the inner radius increases rapidly. The optimum profile was computed to be when $L_1/L_2 = 0{\cdot}4$. The reduced mode conversion is due to phase concellation between the HE_{11} and HE_{12} modes in the large diameter horn section. The effect is relatively broadband because it depends mainly on profile shape.

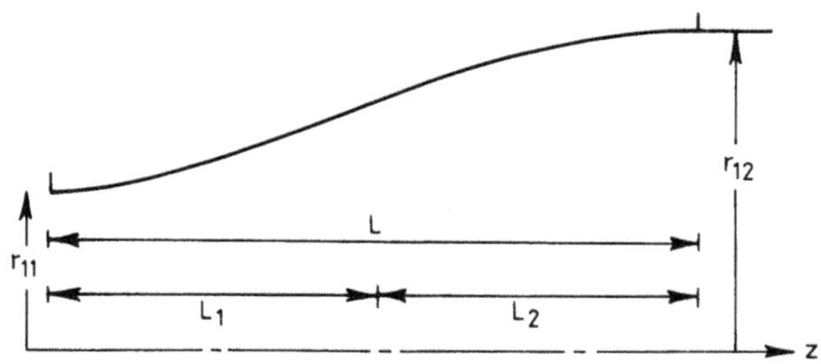

Fig. 5.34 *Geometry of profiled corrugated horn*

5.5 Throat region of horn

Corrugated feeds are usually connected to a section of smooth-wall circular waveguide as a means of transferring power between the radiating horn and the transmission line where signal processing may be conveniently undertaken. The junction is one of the most difficult parts of the feed to design correctly for optimum performance, particularly for wide band operation. Any mismatch between the modes in the circular waveguide and the modes in the throat section of the horn will give rise to mode conversion to higher order modes. Once excited, an unwanted mode may radiate and cause increased cross-polar power and/or reduced efficiency. The effects can become complicated if operation over an appreciable band is required.

We have already studied theoretically (Section 3.6 and 4.4.2) the techniques for analysing the match between a smooth wall waveguide and a corrugated waveguide. The results from field matching studies support the well-used practice that the first slot should be approximately half wavelength deep so that a short-circuit occurs at the edge of the corrugations, that is, at the inner radius of the corrugated waveguide. This argument, however, already introduces a frequency dependent factor, namely the half wavelength deep slot and thus does not answer the question what is the best design away from the central frequency, nor does it indicate how to alter the depths of the slots from the half wavelength value to the value required for radiation purposes. Detailed computer studies, using field matching, have been performed for the throat section, but there are a large number of parameters and probably no one optimum solution so generalisations are not easy.

In this section we shall present a simple graphical method for designing the junction between a conical corrugated horn and a smooth-wall circular waveguide.

This is based on the relationship between the reflection coefficient and the propagation coefficients in a waveguide. It is not claimed to be a rigorous technique but its value lies both in its simplicity and in the insight it gives into the frequency dependent behaviour of the junction. The technique is based on the relationship between the propagation coefficients β_1 and β_2 in two waveguides on either side of a junction and the reflection coefficient, ρ, of the junction

$$|\rho| = \frac{\beta_1 - \beta_2}{\beta_2 - \beta_2} \tag{5.11}$$

This relationship has been derived in Section 3.6. It is in practice more convenient to deal with the corresponding guide wavelengths so

$$|\rho| = \frac{\lambda_{g2} - \lambda_{g1}}{\lambda_{g2} + \lambda_{g1}} \tag{5.12}$$

This means that a perfect match will be obtained when the guide wavelengths in the two waveguides have the same value and that a knowledge of the guide wavelengths will give an immediate estimate of the junction reflection coefficient.

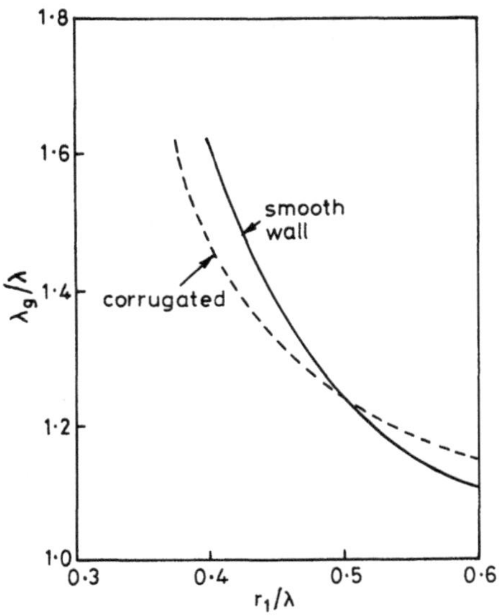

Fig. 5.35 *Normalised guide wavelength of smooth-wall and corrugated waveguides*

We shall concentrate on the problem of matching a TE_{11} mode in a circular waveguide to an HE_{11} mode in a corrugated circular waveguide. The nature of these two modes is such that their guide wavelengths have different dependences on frequency. This is shown in Fig. 5.35 where the guide wavelength is plotted against normalised inner radius (r_1/λ) for a circular waveguide and a corrugated waveguide with slot depth chosen to match at $r_1/\lambda = 0{\cdot}5$. The horizontal scale should be interpreted as changing frequency, not changing radius. The variation with frequency is clearly different for the two waveguides, so a good match can be achieved at only one frequency. The design aim is to ensure that the matched frequency is chosen to give acceptable impedance values over the operating band. If low VSWR is an important specification then Fig. 5.35, in conjunction with eqn. 5.12, tells us that the frequency for perfect match should be nearer the lower end of the frequency band because the rate of change of guide wavelength is greater at lower frequencies. This will mean that the slots will be greater than half a wavelength deep at the upper frequency, an undesirable situation but one which cannot be avoided in a wide band, low VSWR corrugated horn.

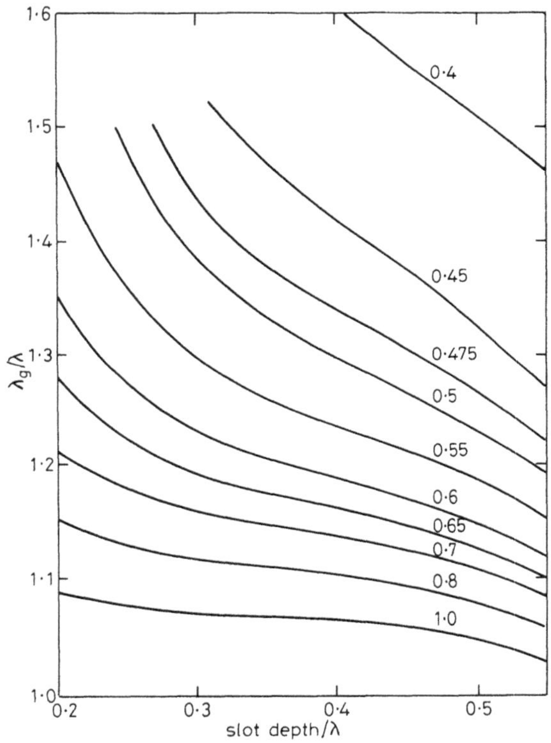

Fig. 5.36 *Normalised guide wavelength against normalised slot depth for corrugated waveguide. Parameter r_1/λ*

The above discussion referred to the junction between two waveguides, whereas in reality we have the junction between one waveguide and a conical corrugated section. This situation can be simulated by approximating the horn as a series of constant diameter corrugated waveguides. The guide wavelength in each corrugated waveguide section can be computed and plotted on a composite graph to give an indication of the way in which the field changes as it propagates along the horn. The guide wavelength will be high at the junction and decrease toward the free-space wavelength at the horn aperture. Sharp changes in guide wavelength along the throat section of the horn are indicative of a change in impedance that can cause a higher order mode to be excited. Thus not only should the guide wavelength on either side of the circular to corrugated waveguide junction be equal, but the change in guide wavelength with distance along the horn, i.e. $\partial\lambda_g/\partial z$ should be as low as possible.

The information needed to design or analyse the throat section of the corrugated horn can be conveniently displayed on a single graph. Curves showing the normalised guide wavelength against normalised slot depth for various normalised inner radii are shown in Fig. 5.36. These have been computed using the space harmonic representation for the fields with a slot width of 0·1λ and a ridge width of 0·05λ. However, in practice the longitudinal slot geometry has only a small influence on the propagation characteristics so the curves can be used for most longitudinal slot geometries.

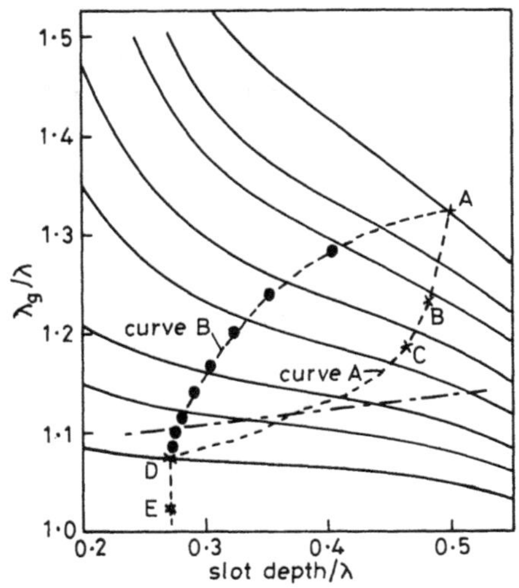

Fig. 5.37 *Trajectories of typical junction*
— · — HE_{12} cut-off

The use of the curves is indicated in Fig. 5.37, curve A. This shows a trajectory for a corrugated horn obtained by plotting the inner radius and slot depth for each corrugation along the horn. The guide wavelength for the smooth-wall waveguide occurs on the slot depth = 0·5λ line, point A. The first slot (from the junction) in the corrugated section gives point B. The second slot point C, and so on. When the tapering of the slots has stopped and the slot depth is constant, point D is reached and from then on the guide wavelength will decrease along the line E.

A good match at the design frequency would have points A and B coincident and a smooth change along the horn. In the example shown in Fig. 5.37, there would be a mismatch at the junction giving a return loss of about −28 dB. The mode conversion along the horn depends on whether a higher order mode is propagating at the relevant radius. The most troublesome mode is the HE_{12} mode because this is excited by changes in waveguide cross-section. The cut-off line for the HE_{12} mode is shown in Fig. 5.37, and in this case the uneven curve occurs at a point in the horn where the HE_{12} mode cannot propagate.

To use these curves to design a throat section, an ideal smooth trajectory would be drawn on Fig. 5.36, between the chosen smooth-wall waveguide value and the chosen final slot-depth value. Points would then be chosen along the curve at which the slots could be placed. Care must be taken to ensure that the inner radius changes smoothly with distance along the horn. A typical design is shown in Fig. 5.37, curve B, with values in Table 5.4. This design has ten slots between the throat ($s = 0{\cdot}5\lambda$) and the final, constant, slot depth ($s = 0{\cdot}27\lambda$) for a horn with a constant flare angle. Ten is a safe number of slots although horns have worked well with as few as four slots. The figure of ten slots is supported by exact computations performed by James [77]. Note that the slot depths shown in Table 5.4 do not decrease linearly with distance. The curve for a linear decrease would be more like curve A in Fig. 5.37, which gives a worse VSWR at the junction. The best match is obtained with a non-linear profile to the horn. The profile can be partly designed with a low VSWR as a design goal using the above procedure. A number of corrugated horns have been designed using the curves and have performed satisfactorily in practice.

Table 5.4 *Slot depths in throat region*

Slot number	Inner radius (wavelengths)	Slot depth (wavelengths)
1	0·45	0·500
2	0·51	0·410
3	0·57	0·360
4	0·63	0·330
5	0·69	0·310
6	0·76	0·290
7	0·82	0·283
8	0·88	0·278
9	0·94	0·273
10	1·0	0·270

A few further points can be deduced from the curves. A better match will be obtained if the smooth-wall waveguide has a normalised radius greater than $0{\cdot}5\lambda$. This is not always possible, but too often a good match is sacrificed because the corrugated horn designer has specified a circular waveguide which is too small in diameter. The better match is obtained because the guide wavelength changes more rapidly for small corrugated or smooth-wall waveguide diameter. A linear taper on the horn does not necessarily produce a good throat section. Particularly in the region of the junction some non-linear taper may give a better match.

The design problem becomes more complicated when a band of frequencies is desired. Then the trajectories at the upper and lower operating frequencies must be placed on the guide wavelength curves. Some compromise will be needed, as a perfect match is possible at only one frequency. The situation is illustrated in Fig. 5.38. Clearly a bad mismatch will occur at the lower frequency, and this can only be rectified by making all the slots slightly deeper so that the design frequency, f_0, is shifted towards the lower end of the operating band. This will improve the lower frequency mismatch, but make the higher frequency mismatch slightly worse. However, since the guide wavelength changes more rapidly at the lower frequencies, the overall effect will be an improved VSWR.

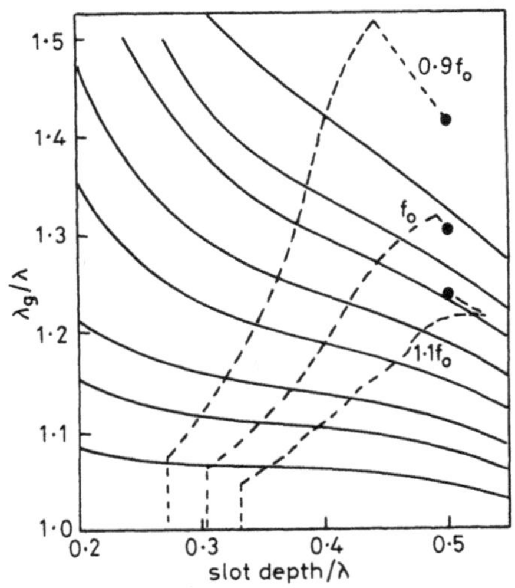

Fig. 5.38 *Trajectories of three frequencies at a junction*

As already mentioned, the longitudinal geometry of the slots has only a marginal influence on mismatch behaviour. It has more influence on the mode conversion along the horn, and in practice we have found it to be better to have a large number of slots per wavelength in the vicinity of the throat. The exact number of slots per wavelength will depend on the application and operating frequency. A value which is ideal, say ten, will be impractical at millimetre wavelengths. James [74, 75, 77] has studied the junction between a TE_{11} mode waveguide and an HE_{11} mode corrugated waveguide (see Section 3.6). His conclusions confirm our discussion. He has designed in detail a throat region and has extended the work to study ring-loaded slots, Fig. 5.39, as a means of increasing the bandwidth of the throat region. He found, theoretically and experimentally, that the return loss could be made to exceed −30 dB over a 1·45 bandwidth ratio with conventional slots. By

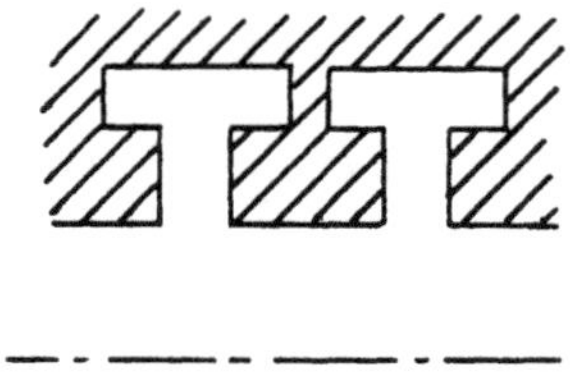

Fig. 5.39 *Ring-loaded slots*

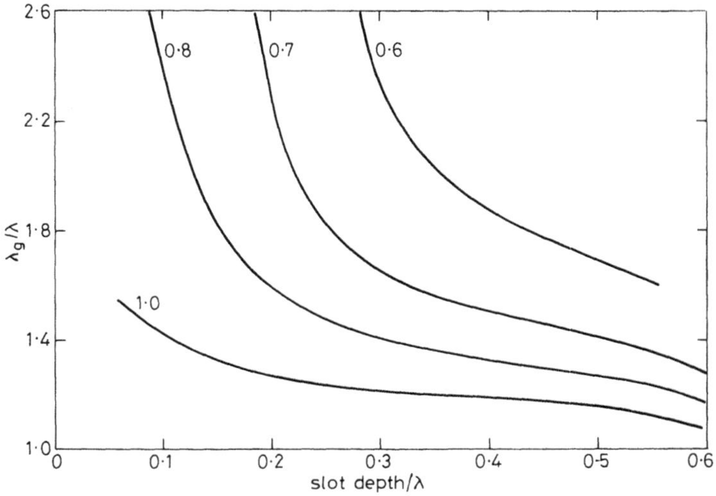

Fig. 5.40 *Normalised guide wavelength against normalised slot depth for HE_{21} mode in corrugated waveguide. Parameter r_1/λ*

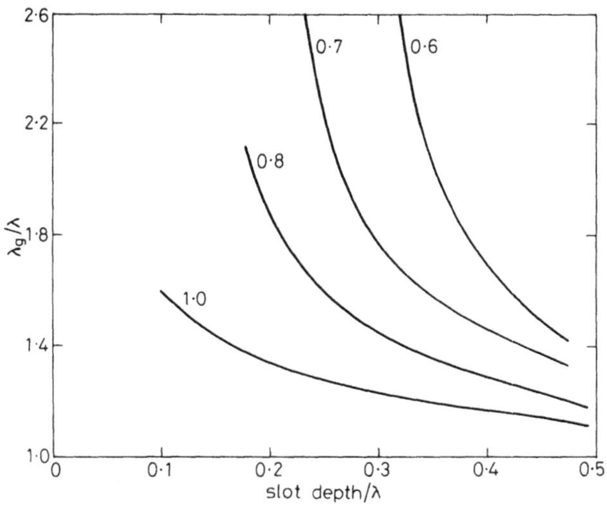

Fig. 5.41 *Normalised guide wavelength against normalised slot depth for E_{02} mode in corrugated waveguide. Parameter r_1/λ*

using ring loaded slots [75] a bandwidth ratio of 1·55 was possible, but this could be increased to at least 2·0 if some HE_{12} higher order mode was accepted. These results are impressive, unfortunately the increased manufacturing complexity and the limitations of the waveguide components probably make it impractical except for a few cases.

There is one further complication to deal with before we complete this section. Sometimes it is required to pass tracking modes (HE_{21}, E_{02}) through the junction as well as the signal mode (HE_{11}). This necessarily implies that the smooth-wall circular waveguide diameter will be larger enough to accommodate the tracking modes. This will help reduce the mismatch on the HE_{11} mode but a perfect match is not possible for all modes. Guide wavelength design curves similar to Fig. 5.36 are shown in Figs. 5.40 and 5.41, for the HE_{21} and E_{02} modes, respectively. The design procedure can be repeated for these modes and a compromise arrived at between the matching of the signal mode and the matching of the tracking modes.

5.6 Gain and directivity

The power gain of any antenna in a specified direction (θ, ϕ) is

$$G(\theta, \phi) = \frac{4\pi \text{ power radiated per unit angle in direction } \theta, \phi}{\text{total power accepted from source}} \tag{5.13}$$

This quantity includes the ohmic or dissipative losses arising from the conductivity of the metal. It is therefore difficult to evalute theoretically and the quantity normally calculated is the directivity. This has a definition similar to gain but the denominator is replaced by 'total power radiated by the antenna'. The peak directivity is normally evaluated, which, for a horn, implies the directivity along the boresight. The peak directivity can be expressed in terms of the tangential aperture electric fields, E_t, of the horn as:

$$D_0(0, 0) = \frac{4\pi}{\lambda^2} \frac{|\int E_t \, dA|^2}{\int |E_t|^2 \, dA} \tag{5.14}$$

where the area A and the fields are expressed in either cylindrical or spherical co-ordinates for open-ended waveguides or conical horns, respectively.

The tangential field expressions inserted into the above equation can be one of the simplified forms discussed in Chapter 3. It is unnecessary to use a precise model unless the horn is a long way from the balanced hybrid condition since the cross-polar components of the horn aperture field do not contribute to the directivity.

The peak directivity has been computed for conical corrugated horns using a spherical wave expansion method, Chapter 4. The results for a wide range of aperture diameters and horn semi-flare angles are shown in Fig. 5.42. Also shown is the directivity for a uniformly illuminated circular aperture, which indicates that the corrugated horn is a relatively inefficient radiator, especially for wide angle horns. As the aperture diameter of the horn is increased, the directivity reaches a maximum value and then oscillates about a lower value. The maximum directivity occurs when

$$\frac{D}{\lambda} \approx \frac{157}{\theta_0} \tag{5.15}$$

and has a value

$$D_0(0,0) \approx 20\log_{10}\left(\frac{250}{\theta_0}\right)\text{dB} \tag{5.16}$$

In both these equations θ_0 is in degrees and they are accurate to within 2% for $5° < \theta_0 < 70°$. The formulae are mainly of use for calculating the gain of wide flare angle horns, since the peak directivity of narrow flare angle horns occur at very large aperture diameters.

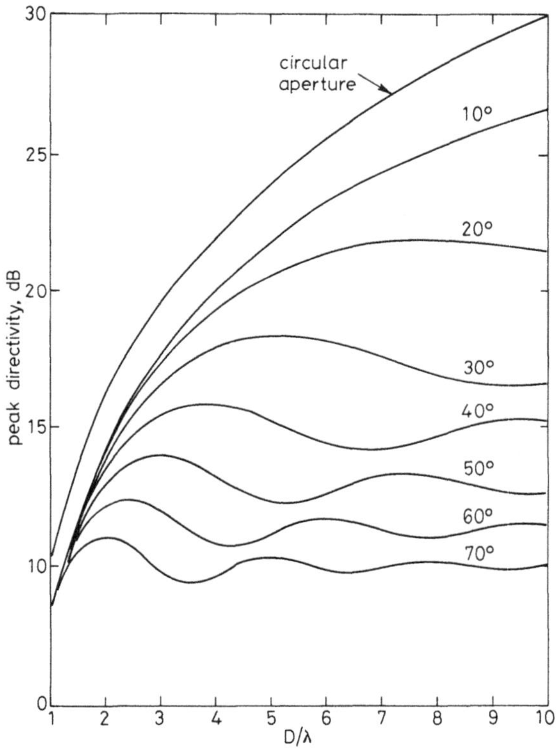

Fig. 5.42 *Directivity against normalised aperture diameter*

The corrugated horn can be used as a standard gain horn [22]. In some respects it is better than the more usual pyramidal horn because the pyramidal horn has high sidelobes and its E plane edge diffraction causes uncertainity in the computed gain.

Very good agreement between the calculated and measured gains of corrugated horns have been recorded. For instance, Chu and Legg [24] obtained agreement of 0·2 dB for a 100 GHz corrugated horn, a difference between theory and experiment, which they attributed to the ohmic losses along the horn.

5.7 Efficiency when used as a feed for a reflector

Most corrugated horns are used as feed for reflectors. The user and designer is interested in the efficiency of the feed as an illuminator of the reflector, that is, a measure of how closely the feed approaches the ideal situation of uniform illumination across the reflector and zero illumination outside the solid angular region subtended by the reflector at the feed focus. In this section we shall consider those components of the efficiency which directly attributable to the feed horn. For most reflector antennas, the feed efficiency components are the main factors deteriorating the overall efficiency, but other components (reflector surface tolerance, aperture blocking, sub-reflector scattering and alignment errors) must not be forgotten.

The gain of a reflector antenna is given by

$$G = \left(\frac{\pi D}{\lambda}\right)^2 \eta \tag{5.17}$$

where D is the diameter of the reflector and η is the total efficiency. This is the product of a number of separate components. The feed efficiency components are identified in Table 2.1, for a paraboloidal reflector. In that table the feed is assumed to possess a radiation function of the form

$$E = \frac{e^{-jkp}}{p} \left(F_E \sin\phi + F_H \cos\phi\right) \tag{5.18}$$

The illumination efficiency, η_I, is a measure of the effectiveness of the feed to accept (or match in reception) the uniformly distributed incident energy in the reflector aperture. It is a function of the amplitude of the feed pattern, and accounts for the different path lengths from the feed to the reflector surface at different angles θ, the space loss. It is a function of the angle θ_0, subtended by the edge of the reflector. This dependence on θ_0 implies that the efficiency of a narrow subtended angle reflector (high f/D) will be greater than a wide subtended angle reflector. Hence, a large aperture horn will give a higher illumination efficiency than a small aperture horn. An ideal feed with $\eta_I = 100\%$ would need to have a feed pattern such that

$$\tfrac{1}{2}(|F_E| + |F_H|) = \sec^2\left(\frac{\theta}{2}\right) \tag{5.19}$$

to yield uniform amplitude in the aperture after optical reflection from the paraboloid. This pattern increases in amplitude as θ is increased, the opposite behaviour of a single hybrid mode feed pattern. Hence, the illumination efficiency of a single mode horn can never approach 100%. There are two ways of improving the illumination efficiency, either by using a feed which propagates one or more higher order modes, or by shaping the reflectors in dual reflectors antennas. The former approach which is discussed in Section 5.8.4 can give a pattern which more nearly approaches the ideal. It has been used in a number of radio-astronomy reflector antennas. The latter approach means that the main reflector will no longer be paraboloidal, and in these circumstances the formulae of Table 2.1 no longer apply. In general, the multimode horn approach is preferred for small reflectors and the latter approach for large, Cassegrain type reflectors.

The spillover efficiency, η_s, is a measure of how much energy spills past the edge of the reflector (or sub-reflector). The wasted energy not only lowers the spillover efficiency but also contributes to raising the antenna noise temperature because the angular regions outside θ_0 'see' background noise. The *spillover* is the energy outside angle θ_0, so spillover $= (1 - \eta_s)$. The feed designer must choose a value of the feed edge taper which gives acceptable spillover. The actual value will depend on the application. In very low noise systems it may be desirable to have a very low spillover with the disadvantage of a reduced illumination efficiency since the outer portions of an unshaped reflector will be underilluminated. In many small earth station antennas, the receiver noise will dominate the total system noise and then the spillover efficiency will be a less important design factor.

The crosspolarisation efficiency, η_x, is a measure of the proportion of the energy which is contained in the the crosspolar radiation pattern. The crosspolar lobes receive unwanted energy in the same way as the sidelobes contribute to the spillover efficiency. In practice the crosspolarisation efficiency can often be taken as 100% because corrugated horns with peak crosspolar levels below -30 dB contribute to a negligible reduction in efficiency.

The last feed efficiency component, the phase efficiency, η_p, is a measure of the phase error across the reflector caused by non-uniform phase of the feed radiation pattern in the same way as the illumination efficiency is a measure of the non-uniform amplitude pattern. An open-ended waveguide has a theoretical uniform phase pattern across the main beam and so have 100% phase efficiency. A horn with a finite flare angle has a non-uniform phase pattern across the main beam so the phase efficiency is less than 100%.

The efficiency formulae given in Table 2.1 have been used to compute the efficiencies of some corrugated horns when used as feeds for a parabolic reflector with edge angle θ_0. The results are shown in Fig. 5.43 for open-ended corrugated waveguides with aperture diameters of 1, 2 and 4 wavelengths, and in Fig. 5.44 for a 4λ aperture diameter corrugated horn with semi-flare angles of 15° and 30°. In all cases a frequency near balanced hybrid has been chosen so that the crosspolarisation efficiency is effectively 100%. The curves show the general behaviour to be expected when a corrugated feed illuminates a reflector. Except for the 1λ diameter feed, the precise numerical values are not applicable to a practical reflector because the larger feeds would be used in Cassegrain reflectors and the scattering loss from the sub-reflector has not been included in the calculations. The total antenna efficiency would also be degraded by the blockage loss and the surface error loss.

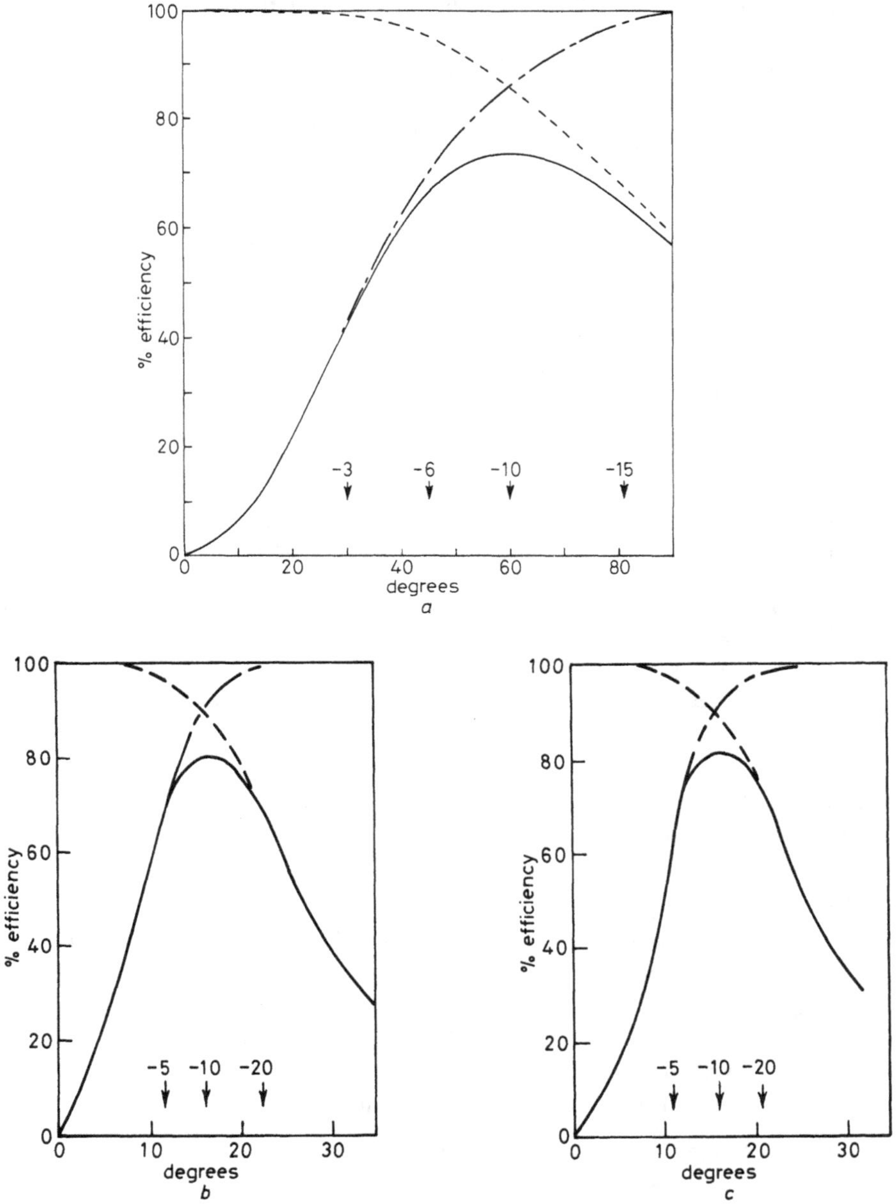

Fig. 5.43 *Efficiency for an open-ended corrugated waveguide. s/λ = 0·27*
——— overall efficiency – – – illumination efficiency – · – spillover efficiency
Arrows indicate copolar power levels. *a* D/λ = 1·0 *b* D/λ = 2·0 *c* D/λ = 4·0

Figs. 5.43 and 5.44 show the compromise between illumination efficiency and spillover efficiency. The form of Fig. 5.43(a) to (c) is similar apart from the increased efficiency as the aperture size is increased. When phase loss is also present, Fig. 5.44 shows that the additional loss in efficiency can be quite significant, which means that a wide flare angle horn is, in general, not an efficient illuminator of a parabolic reflector. The copolar power level is shown in the figure, the peak efficiency in all cases occurs for an edge taper of − 10 dB. This should therefore be the value chosen for most reflector antennas unless the spillover leads to unacceptable noise.

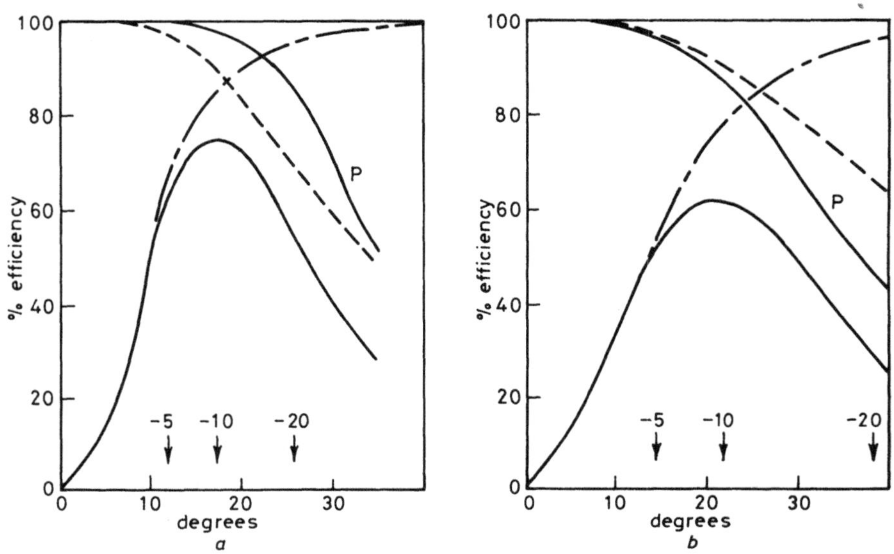

Fig. 5.44 *Efficiencies for corrugated horn. Notation as Fig. 5.43*
a $\theta_1 = 15°$ *b* $\theta_1 = 30°$

G/T parameter

The ratio of gain to noise temperature may sometimes be a more important design parameter than the efficiency alone. This can be computed for a circular reflector by multiplying the efficiency (as calculated above) by $(\pi D/\lambda)^2$ and then by dividing by the noise temperature. If the G/T value is dominated by the 'T' value from the spillover, it may be better to accept a lower edge taper for the feed illumination.

The total noise contribution comes from a number of sources:

(i) the noise received by the main beam of the antenna which will have a minimum level set by the isotropic background radiation level of 2·7 to 3·0 K;
(ii) the spillover noise due to the spillover past the feed;
(iii) the receiver noise.

Only the second component, the spillover noise, is under the control of the feed designer. The spillover noise temperature is given by:

$$T_s = T_a(1 - \eta_s) \tag{5.18}$$

where T_a is the noise temperature 'seen' by the sidelobes of the radiation pattern of the reflector. This will vary according to the location and orientation of antenna. The values of η_s given in Figs. 5.43 and 5.44 show that it usually lies between 0·85 and 0·98. This gives typical values for T_s in the range 1 K to 24 K. Uncooled receivers currently have equivalent noise temperature of 100 K and above. Thus antennas with uncooled receivers will have G/T values dominated by the receiver noise and the spillover need not be the most important design parameter. On the other hand, a cooled receiver may have a noise temperature which is similar to, or less than, the spillover noise. In these circumstances the spillover will determine the G/T value and design of the feed becomes the dominant noise controlling factor.

5.8 Horn design

In the previous sections we have considered each part of the horn and the electrical characteristics separately. Now we shall consider the horn as a whole and discuss the procedure for designing the different types of horn listed in Table 5.1. In most cases this consists of collating the information in previous sections and chapters, but for some special types, e.g. the small aperture diameter horns, additional information is presented. The basic design procedure is similar for all corrugated horns and is enumerated in the following section. It applies to all the types with minor modifications. We shall assume that, unless stated, the corrugated horn is required for use as a feed for a single or multiple reflector antenna. This automatically means that the copolar radiation pattern is the characteristic of principal interest, and it is here that the design of a corrugated horn starts.

Fig. 5.45 *Corrugated horn with semi-flare angle of 8 degrees*

5.8.1 Large aperture, narrow flare angle horns

This type of corrugated horn has seen the widest use as a feed for Cassegrain reflector antennas. The large aperture diameter produces a narrow copolar radiation pattern to illuminate the sub-reflector. The narrow flare angle gives a physically long horn, which is not usually a problem with Cassegrain reflector, and also means that the phase centre of the horn varies less with frequency (see Section 5.4). A sketch of a typical horn is shown in Fig. 5.45. While it is not possible to be precise about what is meant by 'large' and 'narrow', since these are partly tied up with analysis and use, in general the aperture diameter should be greater than 4λ and the semi-flare angle less than 15 degrees. These two criteria mean that approximate methods of analysis for copolar radiation pattern prediction can be used with confidence. The narrow flare angle enables the horn to be treated as an open-ended waveguide which then enables the fields in the waveguide to be expressed as Bessel functions, and the radiation pattern evaluated in closed form. The large diameter also simplifies the analysis and enables further approximations to be made. These have been discussed in Chapter 3 and the equations expressing the radiation characteristics are given by eqns. 3.74–3.81.

The design procedure is as follows:

(i) Choose the aperture diameter. This is done in order to satisfy a specified edge taper for the reflector, or to optimise the efficiency, or to give the highest G/T ratio. The curves of Figs. 5.10–5.12, 5.42 and the discussion of Sections 5.2 and 5.7 can be used for this purpose. The copolar radiation patterns have now been determined and this also fixes the bandwidth for which the crosspolarisation is below a specified level (Section 5.3).

(ii) Choose the geometry of the corrugations. As discussed in Section 5.3 the corrugation geometry determines the frequency at which the minimum level of crosspolarisation occurs. For many applications the crosspolar frequency bandwidth will be much wider than the communication channel bandwidth so the crosspolarisation generated by the horn will be much lower than the specified level. In these circumstances the choice of the corrugation geometry is not critical. In other cases the curves of Figs. 5.22–5.25 can be used to choose the corrugation geometry, and Fig. 5.28 then indicates the bandwidth of the crosspolar characteristics. As a general guideline, for horns with $D/\lambda > 6$, the slot depth can be 0·25λ and the ridge width to slot width ratio should then be between 0·5 and 1·0. The slot dept is measured in the centre of the slot. The number of slots per wavelength will be partly determined by the frequency of operation and the consequent mechanical maufacturing restrictions. At frequencies of a few gigahertz it is possible to contemplate ten slots per wavelength, while at frequencies of hundreds of gigahertz even two slots per wavelength may be difficult to fabricate. We have designed and built horns with a wide variety of a number of slots per wavelength, and we have always found that the horns perform satisfactorily. Similarly, there is no tight constraint in the choice of a ridge width to slot width ratio of $> 0{\cdot}5$. If it is more convenient to make the horn with a smaller ratio, then the only consequence is the need to deepen the slot depth in order to maintain the resonant condition, see Figs. 5.22–5.25. The slot depth also needs to be deeper if the aperture diameter is less than six wavelengths, again see the above figures.

(iii) Choose the flare angle of the horn. Assuming that a linear taper is being used, then the size of the aperture will already place a constraint, thus effectively determining the flare angle. Otherwise the main consideration is the movement of the phase centre as discussed in Section 5.4. If this is not important, perhaps because a relatively narrow bandwidth is required, then choose a semi-flare angle which will reduce the length and hence cost of manufacture, say $10^\circ - 15^\circ$. The patterns of a 15° semi-flare angle horn differ from the open-ended waveguide patterns so some adjustment to the aperture size may be needed. Where the phase centre movement is important a semi-flare angle as small as feasible should be chosen, say 5°.

(iv) Design the junction and throat section of the horn. This was covered in Section 5.5 and the procedure described in that section can be followed. If the ability exists to choose the diameter of the smooth wall waveguide then the matching problem is made easier by choosing a diameter which is well away from the cut-off of the TE_{11} mode, but not large enough to excite high order unity azimuthal dependent modes.

Attentuation: A long corrugated horn will have some loss of power due to attenuation along the horn. This is of particular concern at millimetre and sub-millimetre wavelengths, partly because the loss in the metal increases with frequency, and partly because it is practical to make a horn which is long in wavelengths. The attenuation characteristics have been dealt with in Section 3.4 and are considered in detail in references 39 and 40. The corrugated waveguide propagating a balanced hybrid HE_{11} mode has inherently low attenuation characteristics. This means that there are very few cases where the attenuation is likely to be significant for the corrugated horn user. This is confirmed by measurements carried out at 100 GHz by Chu and Legg [24]. They measured an attenuation of only 0·2 dB due to a 76λ long corrugated horn.

5.8.2 Small aperture horns

Small aperture horns, with a narrow flare angle, are effectively open-ended corrugated waveguides. They can be used as prime feeds for reflector antennas; an example is the antennas used on the European OTS satellite. The design procedure is identical with that for large aperture horns, set out in the previous section. However, in this case, the characteristics are very dependent on the horn dimensions, in particular the crosspolar characteristics will have a narrow bandwidth, as is evident from Fig. 5.27. The slot depth required for resonance will be quite deep as Figs. 5.22–5.25 indicate. This does, however, mean that the matching of the corrugated horn to a smooth-wall waveguide is made easier because the difference between the half wavelength deep slot required for a match and the resonant depth is less than for a large aperture horn.

The characteristics of the small aperture horn are very dependent on the influence of the flange, since the flange is a significant proportion of the total area of the aperture. This has consequences both for analysis and performance. The method used for predicting the radiation patterns must include the effect of the corrugations and the effect of the flange. The Fourier Transform technique should be used since this assumes an aperture is an infinite ground plane, which is, in most cases, a good approximation to the finite flange. The currents on the finite flange have decayed to a low level by the time they reach the outside edge of the flanges. In cases where this is not true, say for feeds with diameters of under 1.4λ, the effect of the finite flange can be taken into account by adding a diffracted term to the radiation from the aperture. The aperture fields cannot, in general, be approximated by simple formulae and it is necessary to take account of the space harmonic to obtain a reliable estimate of the crosspolar levels.

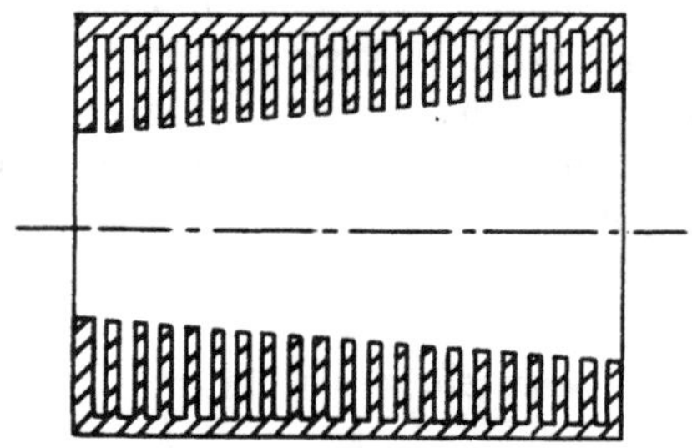

Fig. 5.46 *Small diameter corrugated horn (see text for dimensions)*

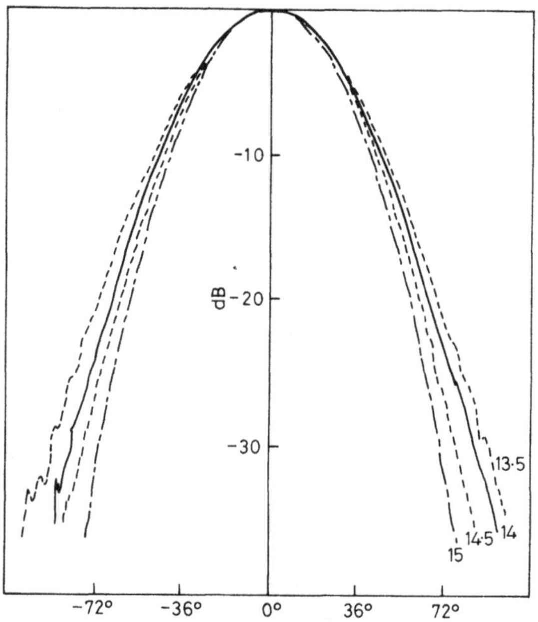

Fig. 5.47 *Measured copolar radiation patterns of horn of Fig. 5.46. Parameter: frequency, GHz*

The cross-section of a horn with an aperture diameter of 1·36λ at the design frequency of 14·25 GHz is shown in Fig. 5.46.

The horn is 3·25λ in length and the semi-flare angle is 5·4°. The figure is drawn to scale. The slot depth is 0·39λ at the aperture and tapers linearly to the throat.

Fig. 5.47 shows the measured E plane patterns over a range of frequencies and Fig. 5.48 shows the measured and computed (using full space harmonic model) 45° plane crosspolar patterns.

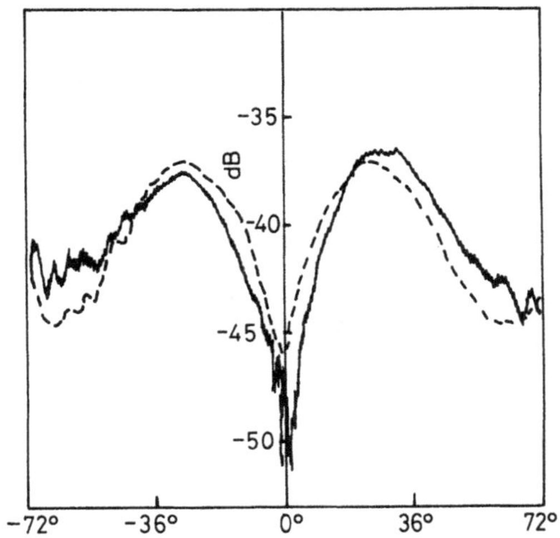

Fig. 5.48 *Measured and computed crosspolar radiation pattern at 14·0 GHz for horn of Fig. 5.46*
— — — computed

The minimum peak crosspolar level was measured as −37 dB, a level set by the radiated contribution from the edge of the flange.

A way to reduce this level is to add a small quater-wavelength deep slot, or choke, in the face of the flange [68, 69, 71, 125] so that the current flow is interruted. Because of the need to limit the flange thickness the choke is re-entrant as shown in Fig. 5.49. This figure also shows the effect of the choke on the crosspolar pattern of a 1·7λ diameter open-ended corrugated waveguide. The crosspolar level has been reduced by 8 dB. The theoretical results shown in Fig. 5.49 are due to Hockham [71] and have been obtained with a matrix solution relating the fields in the aperture to the field in the choke slot. An alternative technique is given by Thomas [144, 149] who extended the corrugations around the corner onto the flange.

5.8.3 Wide flare angle horns

Wide flare angle horns have copolar patterns which are determined mainly by flare angle rather than aperture size. Apart from this fact, the design procedure is similar to the two previous types. The beamwidth curves of Figs. 5.10–5.12 can be used to determine the horn semi-flare angle and the horn length. Although it is possible to design a wide band horn, the boresight directivity is a maximum at the minimum half beamwidth which occurs at a relatively small aperture size where the constant beamwidth does not exist. The peak boresight directivity enables an optimum size to be determined for any semi-flare angle, eqn. 5.15. This is shown in Fig. 5.50 and the corresponding −3 dB, −10 dB and −20 dB half beamwidths in Fig. 5.51. Note that the −15 dB half beamwidth level is approximately equal to the semi-flare angle of the horn.

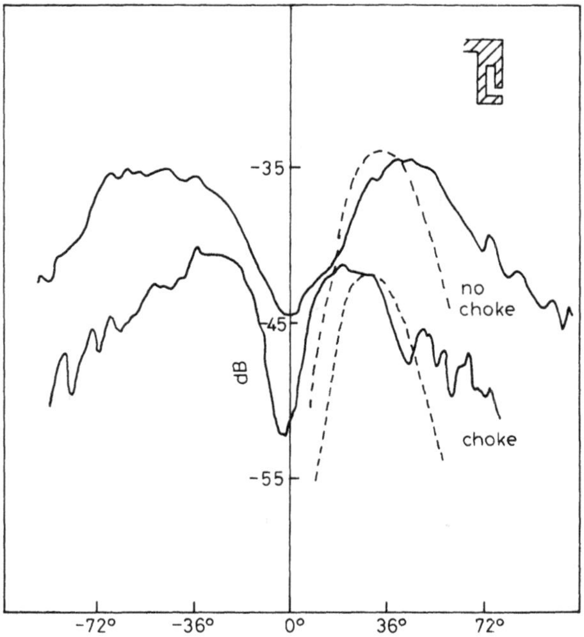

Fig. 5.49 *Effect of choke in flange on the crosspolar radiation pattern of a 1.7λ diameter corrugated horn*
——— measured – – – theory

The corrugation geometry is designed in the same way as for large aperture corrugated horns, namely by choosing the slot depth according to the radial slot geometry. The crosspolar characteristics are frequency dependent so the wide band properties will only apply to the peak crosspolar level if the aperture size is large in wavelengths. It is normal with wide flare angle horns to cut the slots perpendicular to the wall of the horn rather than perpendicular to the z axis, but this does not modify the design procedure.

The phase centre in a wide flare angle horn will remain essentially fixed at the apex. This can be an advantage since, unlike the narrow flare angle horns, the phase centre will remain fixed as frequency changes. A comparision of efficiencies shows that the wide flare angle horn has a lower efficiency than the narrow angle horn but if the moving phase centre is taken into account, the wide flare angle horn can perform better as frequency is increased above the design frequency.

The throat section of the horn is crucial because the change from a smooth wall waveguide to the conical section can be very abrupt and thus may give rise to higher order modes which will considerably distort the radiation patterns (this is discussed

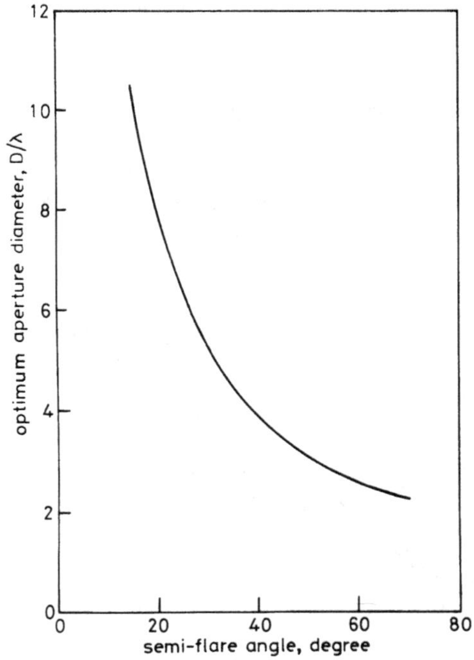

Fig. 5.50 *Optimum normalised aperture diameter against semi-flare angle for wide angle corrugated horn*

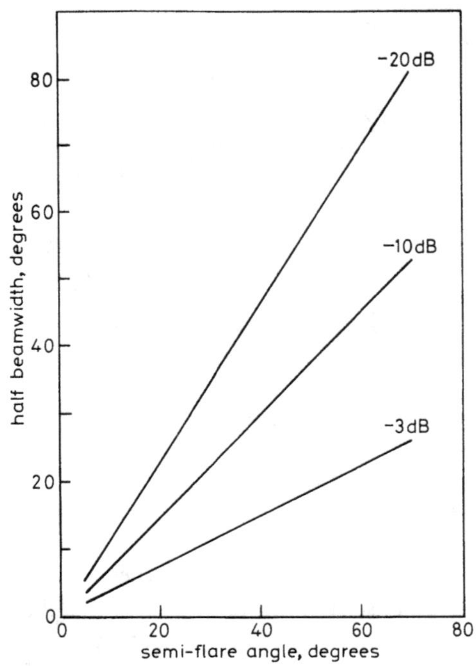

Fig. 5.51 *Half beamwidths corresponding to Fig. 5.50*

in detail in Section 4.4). We have found that the design of the throat section can be performed using the procedure of Section 5.5. This will normally mean that a small section of smooth-wall conical horn will be needed between the smooth-wall waveguide and the conical corrugated section. Fig. 5.52 shows the cross-section of a 30° semi-flare angle horn with very low mode conversion properties. This is confirmed by the measured peak crosspolar characteristics, Fig. 5.53, which follow the theoretical prediction (using a spherical wave expansion method, Chapter 4) very well. Note the low level of crosspolarisation over a appreciable band of frequencies.

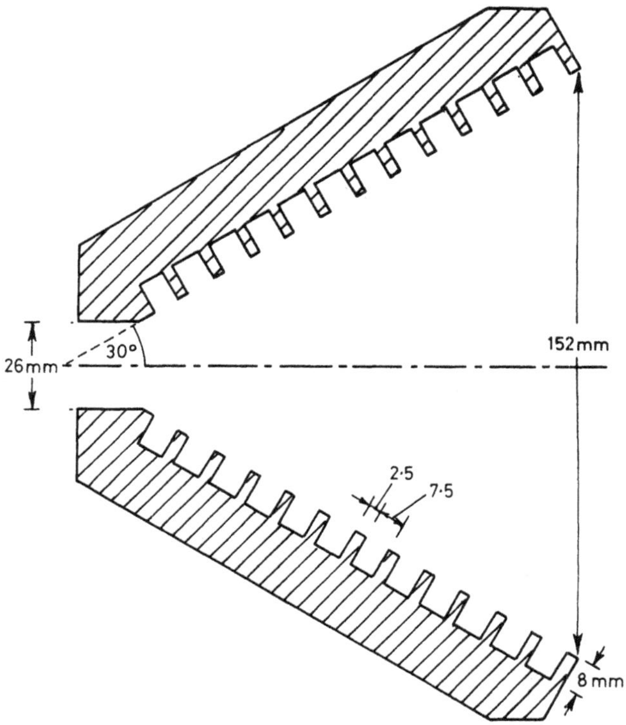

Fig. 5.52 *Scale diagram of 30 degree semi-flare angle corrugated horn*

The slots are of constant depth along the horn. This may appear to be in contradiction to the previous discussions, but if the inner radius of the first slot (0·55λ) and the slot depth (0·27λ) are plotted on Fig. 5.36, we can see that it has a guide wavelength nearly identical to the smooth-wall waveguide value ($\lambda_g/\lambda = 1{\cdot}4$). The measured return loss at the design frequency is −30 dB.

Wide flare angle horns can be made to work when they are very small in wavelengths. This has been demonstrated theoretically and experimentally by Estin *et al.* [61] who used a 45° semi-flare angle horn with an aperture diameter of between 1·2λ and 2·8λ. The peak crosspolar level at midband was −30 dB, a very good level

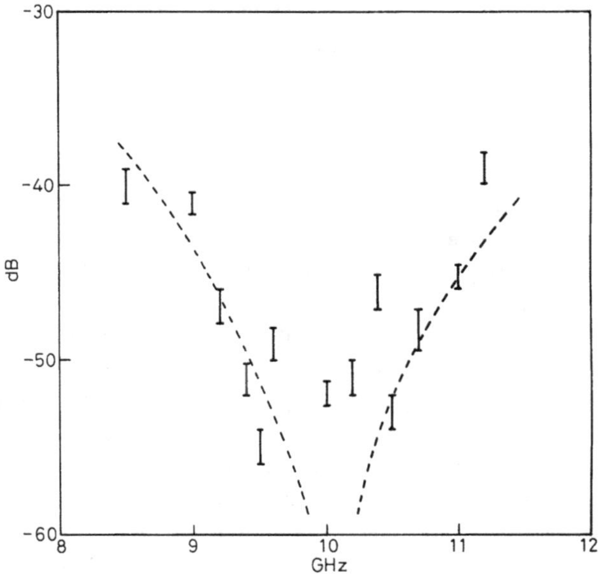

Fig. 5.53 *Measured peak crosspolar power against frequency of 30° horn of Fig. 5.51*
— — — theory

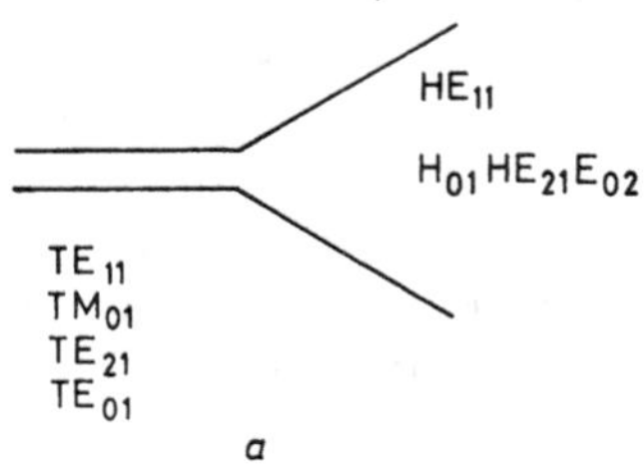

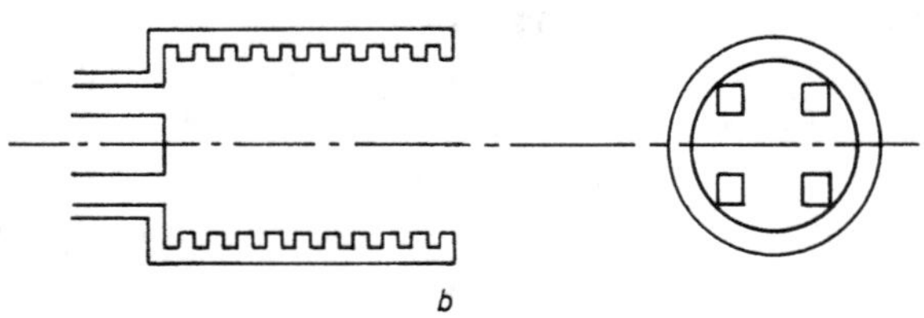

Fig. 5.54 *Multimode corrugated horns*
a Horn with tracking modes *b* Monopulse corrugated waveguide with four feeder waveguides

for such a small horn. The horn was investigated for use as a gain standard at low microwave frequencies, when the physical size of an antenna becomes the limiting design factor.

5.8.4 Multimode horns

Tracking horns: Tracking horns use a number of modes to provide tracking or monopulse information, which may be in addition to a signal mode or it may be the only function. Thus a primary requirement is that the horn should support all the modes throughout the length of the horn. This is clearly no problem at the aperture but means that the minimum size of the throat region is determined by the mode with the highest cut-off frequency. A major design limitation with these types of horn is the need to separate the power from the various modes into the tracking channels. The separation can be done in two ways. Either the horn has a cross-section similar to a standard horn with the modes being transformed from the corrugated section to the corresponding modes in smooth-wall rectangular waveguide (Fig. 5.54(a)), and the signal extraction is then done in smooth-wall waveguide. Or the end of the corrugated horn is terminated abruptly and a number of smooth-wall waveguides butt directly on to the corrugated section (Fig.5.54(b)). The smooth-wall waveguides then carry the tracking modes. An example of the former approach is Watson *et al.* [161], where a sophisticated mode extraction network is described, and an example of the latter approach is Clarricoats and Elliot [25]. Watson *et al.* needed a corrugated horn which had a primary function of transmitting or receiving a communication channel so the design of the horn followed standard procedure except for the need for an oversized throat region. In contrast, the horn described by Clarricoats and Elliot consisted of a length of corrugated waveguide with no flare. The end was terminated directly in a four horn cluster. Because there was no significant bandwidth requirement this was the easiest approach. The length of the waveguide and the inner and outer diameters are chosen to propagate the modes in the correct phase relationship. The principles of this type of horn have been covered in Chapter 3.

A modification to the four horn cluster feed is to add a fifth horn [50, 154]. This improves the tracking performance but involves some compromise in the corrugated horn design. A thorough investigation has been carried out by Viggh [154].

The design procedure for a tracking horn follows the same principles as enumerated for large aperture horns, except that the design of the copolar and crosspolar patterns are replaced by the design of the modes required for tracking. This means an aperture field which radiates a null on boresight. Since the essential feature is the null, the level and position of the peak on either side of boresight is not normally of concern, although the slope of the difference pattern with angle is important. The form of the radiation patterns of the HE_{21}, E_{02} and H_{01} modes are shown in Fig. 5.16. These have normalised to 0 dB, but the shape of all the modes is seen to be nearly identical.

The throat section can be used as a filter of unwanted modes by choosing the inner diameter to just propagate the tracking modes. It is desirable for the tracking modes to have the same differential phase shift as they propagate down the horn. This implies that the dispersion characteristics should be nearly coincident. An example is shown in Fig. 5.55 where the E_{02} and HE_{21} modes have similar propagation coefficiencies over a range of frequencies. Design of the tracking horns is best accomplished for the particular specification by using a computer program. The simple waveguide model is adequate for this purpose. The corrugation geometry is normally chosen to be resonant for the HE_{11} mode at the design frequency, even though the higher order modes may have a different resonant depth.

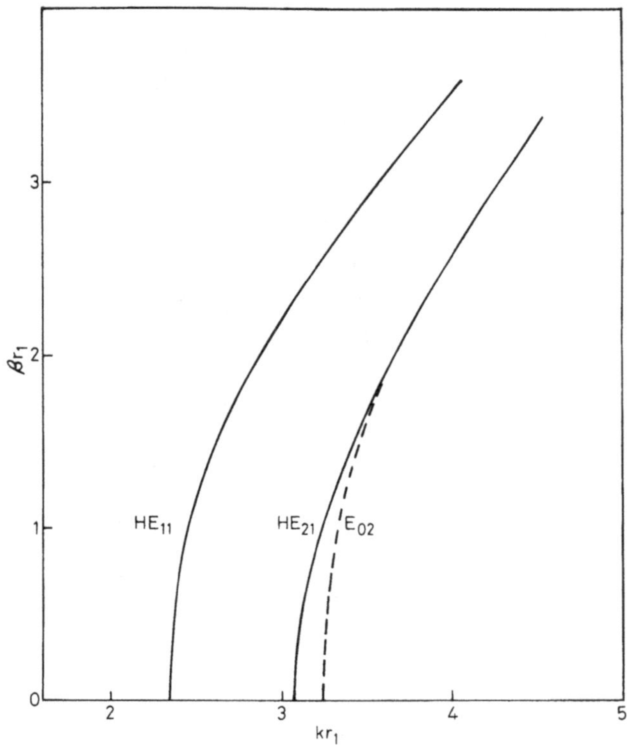

Fig. 5.55 *Dispersion characteristics of HE_{21} and E_{02} modes in corrugated waveguide with $r_1/r_0 = 0{\cdot}6$*

Shaped pattern horns: A class of multimode horns have been used for prime focus paraboloidal reflectors. These produce a small dip on boresight and peak power a few degrees off boresight [132, 157]. The purpose of this pattern is to overcome the space attenuation loss in a paraboloidal reflector and produce nearly uniform illumination over the reflector surface, as illustrated in Fig. 5.56. The pattern can be obtained by exciting a proportion of HE_{12} mode which is added to the main HE_{11} mode. This is normally done in a section of corrugated waveguide abutting

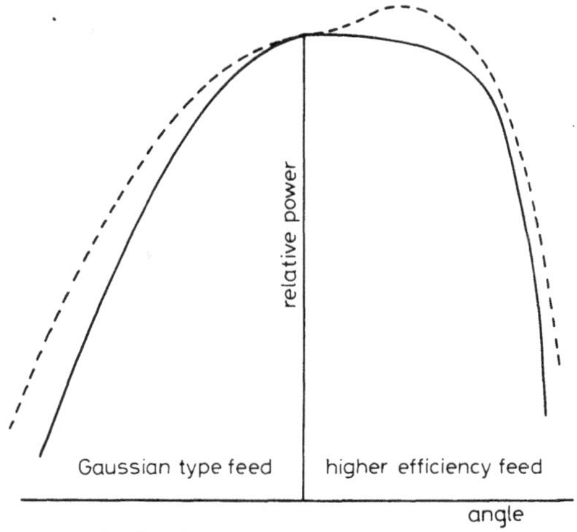

Fig. 5.56 *Feed pattern required to increase aperture efficiency in a prime focus reflector*
——— aperture field – – – field from feed

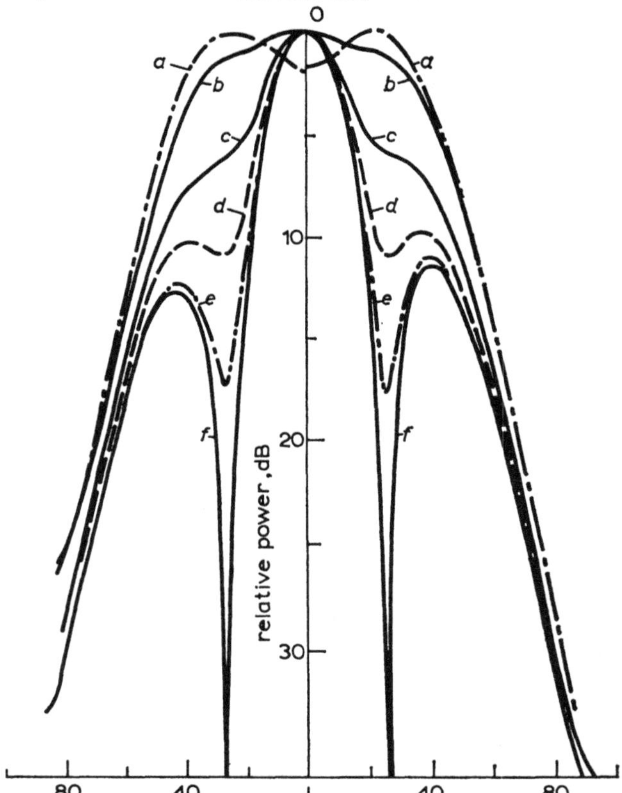

Fig. 5.57 *Patterns of two hybrid mode horn*
Relative phase of modes: *a* – 0° *b* – 36° *c* – 72° *d* – 108° *e* – 144° *f* – 180°

directly onto a smaller diameter smooth-wall circular waveguide. The step excites the HE_{12} mode in the correct proportion. Fig. 5.57 shows the patterns which can be obtained by adding together HE_{11} and HE_{12} mode power in various phases. The pattern marked 'a' is the desired pattern. The junction between the corrugated waveguide and the smooth-wall waveguide should be designed with a modal matching procedure [49].

A dual hybrid mode ($HE_{11} + HE_{12}$) horn is described by Thomas and Bathker [153]. This used a step junction between a smooth-wall circular waveguide and a conical corrugated horn with a semi-flare angle of 6·25°, to generate the HE_{12} mode. The horn exhibited 1% improved spillover efficiency and 5% improvement in illumination efficiency by comparison with the equivalent single mode horn feeding a Cassegrain antenna. Overall the gain of the antenna improved by 0·36 dB.

Matched feeds: Another type of shaped pattern feed is the 'matched feeds' invented by Rudge and Adatia [189]. These are designed to produce a copolar and crosspolar pattern which matches the focal region fields of an offset reflector, so eliminating the inherent crosspolarisation present in an offset paraboloid. Although the matched feeds have been mainly designed using smooth-wall horns, corrugated matched feeds can be produced. A proportion of HE_{21} mode is added to the fundamental HE_{11} mode to give the desired pattern. The theory and design is fully covered in Reference 132.

5.8.5 Broadband horns

Horns which have nearly constant copolar radiation patterns over a very wide band of frequencies can be produced. We should emphasise that the broadband characteristic applies only to the copolar characteristics. The crosspolar characteristics,

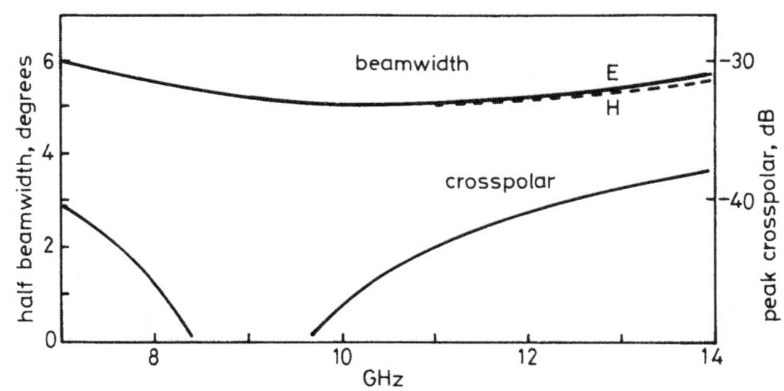

Fig. 5.58 ***Half beamwidth and peak crosspolarisation against frequency of a broadband corrugated horn with 15 degree semi-flare angle and aperture diameter = 262 mm***

being determined by the resonant nature of the corrugations, are necessarily relatively narrow band. Reference to the −3 dB beamwidth curves of Fig. 5.10 will show that a horn with a semi-flare angle of between 10° and 20° has a constant beamwidth over a wide range of D/λ. The optimum semi-flare angle is 15° when the −3 dB half beamwidth is between 5° and 6° from $D/\lambda = 7$ to $D/\lambda = 14$. The E plane, H plane and 45° crosspolar patterns across the band are as shown in Fig. 5.58. Because the aperture is large, the theoretical crosspolarisation is, in fact, low across the whole of the band.

The throat region is a limiting factor for broadband horns. If the first slot is chosen to be half a wavelength deep at the top end of the frequency band then it is found that the HE_{11} mode is cut off at the lowest frequency. Thus the slots must be greater than half a wavelength deep at the upper frequency, a consequence which could give rise to mode conversion and thus a useful range of frequencies less than one octave. We are not aware that corrugated horns have been measured over a very wide range of frequencies, so we have no practical evidence of the limits of operation of corrugated horns.

5.8.6 *Multi-frequency horns*

This type of horn is distinguished from the previous type in that the horn is required to operate at a number of discrete frequencies rather than over a wide band of frequencies. The corrugated horn has the ability to operate whenever the slots are $(2n-1)\lambda/4$ deep ($n = 1, 2\ldots$). An example of a large dual frequency band corrugated horn is provided by the feed for some of the NASA Deep Space Network reflector antennas. This is described by Williams and Withington [163] and operates, simultaneously, from 2·0 GHz to 2·4 GHz (where the slot depth is 0·34λ to 0·4λ) and 7·1 GHz to 8·6 GHz (where the slot depth is 1·21λ to 1·44λ). The semi-flare angle of the horn is 17·1° at the 1·07 m aperture. Thus the copolar patterns are functions of angle, resulting in similar beamwidths at the two operating bands.

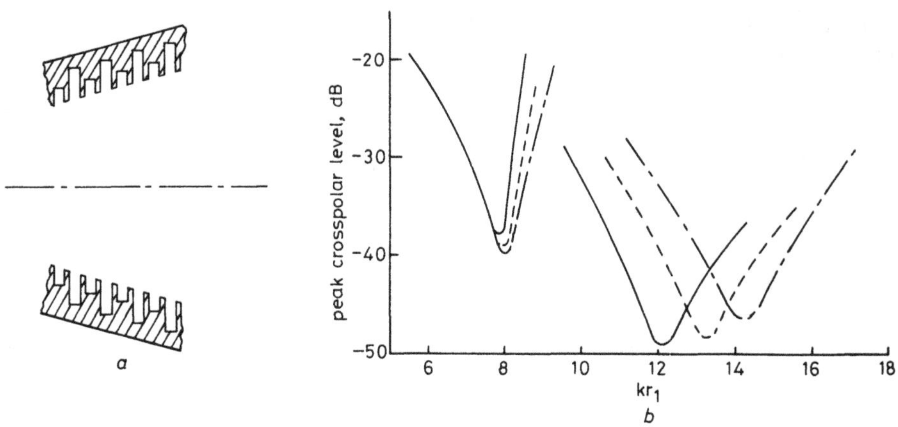

Fig. 5.59 *Dual depth corrugated horn*
a Cross-section *b* Peak crosspolar power $r_1 = 41{\cdot}5$ mm, $b = 3$ mm, $t = 1{\cdot}5$ mm, $s_1 = 8{\cdot}5$ mm, ——— $s_1/s_2 = 1{\cdot}5$, – – – $s_1/s_2 = 1{\cdot}7$, - - - - $s_1/s_2 = 1{\cdot}9$

For many applications the specified channel spacing does not satisfy the above slot depth criteria. An example is the transmit and receive bands for direct broadcast satellites. An alternative approach is to use a dual depth corrugated horn, Fig.

5.59(a) [5, 63, 64, 124]. One of the slot depths is resonant at one frequency, while the adjacent slot is resonant on a second frequency. Thus reasonable performance can be expected at two discrete frequencies. An experimental model has been measured by Ghosh *et al.* [63] who showed that low crosspolarisation could be obtained in two frequency bands, spaced 1:1·5 apart. The theoretical prediction of the crosspolar performance needs to be performed using the space-harmonic model because the influence of the dual depth slots cannot be accurately represented by a surface impedance. We have extended the space harmonic analysis given in the Appendix to the dual-depth case. The characteristic matrix equation which must be solved to determine the propagation relationships is extremely complicated because each term in the matrix is complex. The space harmonics modify the frequency at which minimum crosspolarisation occurs in a manner similar to the results presented in Section 5.3 for the standard corrugated waveguide. Fig. 5.59(b) shows typical results for fixed slot width and ridge width and varying depth of the shallower slot. The ratio of the upper resonant frequency to the lower resonant frequency is seen to correspond to the ratio of the slot depths. The results are for an open-ended corrugated waveguide but will apply also for narrow flare angle corrugated horns.

5.8.7 Compact horns

The compact, or profiled, horn, Fig. 5.60, has already been discussed in Sections

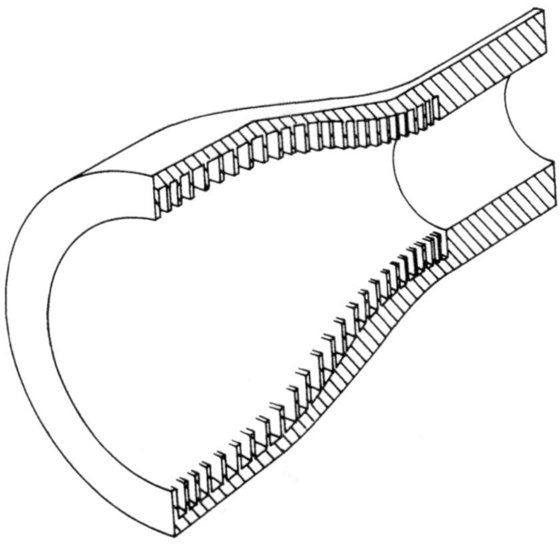

Fig. 5.60 *Profiled corrugated horn*

4.4.4 and 5.4. The design of the horn is straightforward, following the standard procedure laid out for large aperture horns. That is, design the aperture diameter for the copolar pattern using open-ended corrugated waveguide results; then design the aperture corrugation geometry for the crosspolar characteristics. The flare section is designed according to eqn. 5.10 and the throat section is designed according to Section 5.5. It is advisable to allow for a small length of corrugated

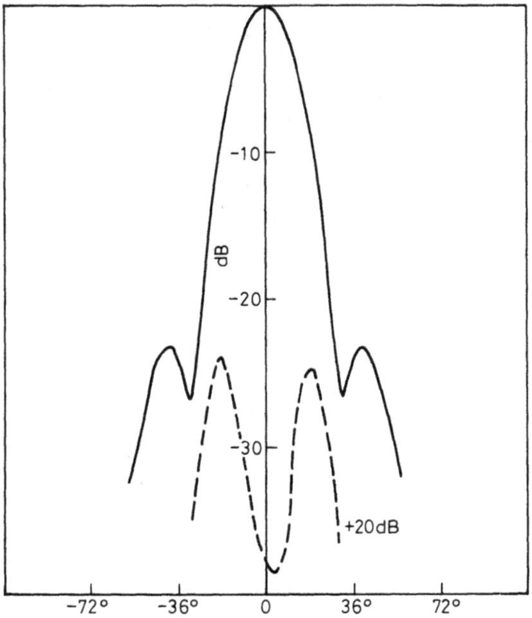

Fig. 5.61 *Measured radiation patterns of profiled horn (see text for dimensions)*
——— copolar – – – crosspolar

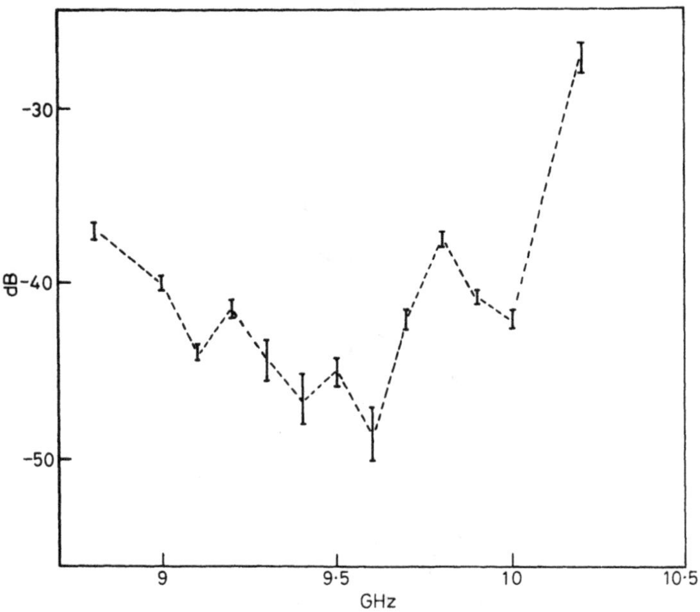

Fig. 5.62 *Measured peak crosspolarisation of profiled horn of Fig. 5.61*

waveguide at the aperture, after the profiled section, to ensure that the phase centre remains fixed at the aperture. Measured copolar and crosspolar patterns of a 2·9λ diameter horn are shown in Figs. 5.61 and 5.62. The horn has an aperture diameter of 89 mm, a slot depth of 8·7 mm, a slot width of 3 mm and a ridge width of 1 mm. At the throat the slot depth is 13·4 mm and the diameter of the smooth-wall waveguide is 33 mm. The total length of the corrugated horn is 129 mm which includes a 17 mm waveguide section at the aperture. The measured peak cross-polar level is seen to be less than −40 dB over a 0·75 GHz band. The excitation of some EH_{12} mode limits the high frequency performance. As discussed earlier this is a consequence of using a non-linear taper. It can also be seen from Fig. 5.61 that the copolar patterns are distorted by the presence of some HE_{12} modes. This causes the sidelobe at 40 degrees. Fortunately it is outside the angle of the beam needed for illuminating a reflector so is not serious.

5.9 Numerical prediction of performance

The design of corrugated horns usually involves a first stage of analysis, then a second stage of numerical prediction of the theoretical performance and finally a third stage of manufacture and testing. In this section we shall collect together and summarise the available methods for numerically predicting the characteristics of corrugated horns. In order to predict the radiation patterns, the propagation coefficient in the circular corrugated waveguide or the conical corrugated wave-guide must first be computed, and then aperture fields predicted. From these fields the near-field or far-field radiation patterns can be computed. The design of the throat section also requires a knowledge of the propagation coefficient.

5.9.1 Propagation characteristics

(a) Corrugated circular waveguides: When the normalised diameter of the corru-gated waveguide is large, and the waveguide is operating near to the balanced hybrid condition the propagation characteristic can be estimated from eqn. 3.37, i.e.

$$J_q(x_1) = 0$$

where $q = 0$ for HE_{1m} modes and $q = 2$ for EH_{1m} modes, and the wavenumber $x_1^2 = (kr_1)^2 - (\beta r_1)^2$. This equation has the advantage that the relationship between kr_1 and βr_1 is in closed form, a situation not normally encountered in inhomo-geneous waveguides. For the modes of interest the values of x_1 are given by:

Mode	x_1
HE_{11}	2·405
EH_{12}	5·135
HE_{12}	5·520
EH_{13}	7·016
HE_{13}	8·654

The equation has a surprisingly wide range of validity. Even for frequencies far from the balanced hybrid condition, the values predicted are often good estimates of βr_1 and kr_1. Dragone [56] has derived a series relationship for x_1 which gives an improved estimate for the wavenumber. The first two terms are

$$x = x_1\left(1 - \frac{Y}{2kr_1}\right)$$

where Y is the admittance of the corrugations. However, the relationship is only valid near balanced hybrid (where $Y = 0$) and for large kr_1 so the second term contributes only a small perturbation to x_1. Higher order terms converge slowly and need using with care.

A more accurate computation of the propagation characteristic involves the solution of a transcendental equation. The formulation based on the surface impedance approach is given by eqn. 3.26. In order to solve this equation for fixed kr_1 trial values of βr_1 must be used and an iterative scheme used to converge on the correct value of βr_1. The equation is valid for all m and both fast and slow wave regions of the propagation characteristic. A sample computer program is given below. The main computational difficulty lies in locating suitable trial values for βr_1, and experience is the only reliable guide as to how to make the choice. Because the propagation equation involves combinations of Bessel functions there can be regions where poles occur and the value of the equation changes very rapidly as the parameters are altered. The solution to this problem is to rewrite the equation in such a way as to remove the pole.

The most accurate method of computing the propagation coefficient in corrugated waveguides is to use the space harmonic formulation given in Appendix. The characteristic equation is given by eqn. A29. A determinant must be evaluated. Each term of which is a combination of Bessel functions. The procedure for solving the propagation coefficients is similar to the surface impedance case with the difference that the computational time is longer due to the increased complexity of each term. The space harmonic formulation is mainly of use for accurate prediction of the aperture field and then of the crosspolar radiation patterns. It is also useful if the corrugated waveguide has non-standard corrugation geometry so that the influence of the slots cannot be represented as an impedance surface.

(b) Corrugated conical waveguides: A closed form solution for the spherical wavenumber, ν, is given by eqn. 4.19, namely:

$$\nu = \frac{2{\cdot}405}{(\theta_1 \sin\theta_1)^{1/2}} - 0{\cdot}5 \qquad (\text{for } HE_{11} \text{ mode})$$

This gives values to within 1% of the true values for semi-flare angles up to 60°. The formula is valid for small flare angles (< 10°) but for these horns it is better to approximate the conical horn as an open-ended waveguide and use the simplier circular waveguide function.

To derive more accurate values for ν, the characteristic eqn. 4.14 must be solved. This is a transcendental equation with Legendre functions, and is solved in the same manner as the characteristic equation for corrugated circular waveguides. See Table 4.1.

(c) Program to compute propagation and attenuation characteristics: The program listed in Program 1 computes the propagation and attenuation characteristics of corrugated circular waveguides by implementing the surface impedance formulation, eqn. 3.26, and then using perturbation formulae equations, eqns. 3.66, 3.68–3.71, to calculate the attenuation in the waveguide. The program is written in BASIC and is suitable for running on many types of personal computers. It computes the value of βr_1 at a fixed frequency. An initial value of βr_1 is either input as data or calculated using $x_1 = 2{\cdot}405$, then the values of characteristic equation are calculated as βr_1 and is incremented until a root is found. Convergence on to the root is done with a Langrangian inverse interpolation procedure.

The input data is specified in Table 5.5 and sample data and results in Table 5.6. The program can be used to compute the values of βr_1 appropriate to the throat region of the horn when the simple approximation of eqn. 3.37 does not apply or when the modes with non-unity azimuthal order are required.

Table 5.5 *Input data for Program 1*

N	=	Azimuthal order
RA	=	Slot depth (metres)
R1	=	Inner radius (metres)
D	=	Slot width (metres) (used for attenuation calculation)
TT	=	Ridge width (metres) (used for attenuation calculation)
RH	=	Conductivity of walls (mhos/m)
FR	=	Frequency (GHz). If FR = − 1 jumps to start of data.
C(1)	=	Initial trial for r_1. If C(1) = − 1 calculate from $x_1 = 2{\cdot}405$.

After computation returns to line 1110. The accuracy of computation is set by line 1290

(d) Bessel functions: The computation of the propagation and radiation characteristics of circular waveguides need Bessel functions. Since most of the computer time is taken up with computing the Bessel function it is important to use an efficient routine. For inhomogeneous waveguides, Bessel functions of the first and second kind (J and Y) are needed and also the corresponding modified functions (I and K). There are two main methods of computing the functions, either using a polynominal approximation or using a recurrence relationship. The polynomial approximation method (e.g. Abramowitz and Stegun [178], Chapter 9) requires a reasonable amount of memory to store the coefficients. It is the method usually implemented in the maths libraries of large computers but is not suited to small computers. The recurrence relationship using backward recursion is an efficient

Table 5.6

```
RUN
? 1,.01,.03,.007,.003,1·557E7
? 10, – 1

Corrugated Waveguide

Azimuthal No.  = 1
R1/R0          = 0·75
R1             = 0·03 metres
Slot depth     = 0·01 metres
Slot Width     = 7E – 03 metres
Ridge Width    = 3E – 03 metres
Conductivity   = 1·557E + 07

Frequency      = 10 GHz
k*r1           = 6·28318
Beta*r1        = 5·84885
Hybrid factor  = 1·22057
Attenuation    = 1·18206E – 04 dB/m
```

method for the typical argument values needed in corrugated horn design. Also, when more than one Bessel function is required, the same formulae can be used to generate all functions. This method is implemented in Program 1, lines 2160 to 2520, which generate $J_m(x)$, $Y_m(x)$, $I_m(x)$, $K_m(x)$ and their derivatives. The code is compact and has the advantage that the accuracy can be changed to suit individual circumstances. This is done by changing the argument in line 2230. The accuracy, as given, is six decimal places.

A compact simplified version of the same routine to calculate $J_0(x)$ and $J_1(x)$ is contained in Program 2, lines 1400 to 1500. This routine is particularly useful for radiation pattern prediction programs.

5.9.2 Radiation characteristics

(a) Open ended waveguides and narrow flare angle horns: The copolar and crosspolar radiation patterns can be predicted either by the Kirchhoff–Huygen method or by the Fourier Transform method. The Fourier Transform method is better because it requires only the transverse electric fields in the aperture of the horn, and because the assumption of an aperture in an infinite ground plane is a good approximation to the finite flange which must be present on corrugated structures. The Fourier Transform method also leads to expressions which are easy to compute for open-ended waveguides They are given in eqns. 3.74–3.81, for the case when space harmonics are neglected. The values of β and ω can be found either by the balanced hybrid formula, eqn. 3.37, or the surface impedance characteristic eqn. 3.26.

The Fourier Transform method can also be applied to the space-harmonic formulation to give analytic expressions for the radiated fields, and these are given in the Appendix. This formulation is needed for an accurate prediction of the cross-polar characteristics when either the corrugation geometry is unusual (e.g. for few slots per wavelength) or the aperture diameter is small so that the slots have to be deep in order to resonate.

When the horn flare angle is significant a spherical phase factor can be added to the aperture fields to improve the prediction of the copolar sidelobe region. This has been studied in Section 4.7 and eqn. 4.112 gives the phase factor. The radiated field components are given by eqn. 3.73 with integrand multiplied by exp $(j\,\phi(r)]$. The radial integral is no longer given in an analytic form and numerical integration is needed to obtain the far-field results. The aperture field of corrugated horns is well behaved so most integration routines (e.g. Simpson's rule) converge rapidly.

(b) Wide flare angle horns: When the semi-flare angle of a horn is greater than about 15° the spherical wave expansion method must be used. The expressions for the radiated fields, are given in eqns. 4.39–4.41. These expressions are valid at any distance R from the aperture and mean that near-field patterns or far-field patterns

Table 5.7

RUN
Radiation Pattern of Corrugated Waveguide

Inner radius in Wavelengths = ? 3·0
Slot depth in Wavelengths = ? 0·28

Slot admittance = − 0·190764
Hybrid factor = 1·03346

Increment angle, Final angle = ? 2·20

ANGLE	COPOLAR (dB)	X-POLAR (dB)
0	0	− 100
2	− 0·290878	− 69·4263
4	− 1·17578	− 58·0885
6	− 2·69489	− 52·2385
8	− 4·92867	− 48·961
10	− 8·02951	− 47·3925
12	− 12·3091	− 47·2265
14	− 18·5556	− 48·4368
16	− 30·2667	− 51·3027
18	− 33·8935	− 56·9102
20	− 27·8657	− 74·0154

can be computed using the spherical Hankel functions and associated Legendre polynomials. Neither function is time consuming to compute but these routines are not normally available in standard computer software libraries.

(c) Computer program for radiation characteristics of narrow flare angle horns: A simple BASIC program to compute the copolar and crosspolar radiation patterns of the HE_{11} mode in narrow flare angle corrugated horns or open-ended corrugated waveguides is listed in Program 2. This is based on eqns. 3.80 and 3.81 with the wavenumber computed using the first two terms of eqn. 3.39. The input data is self-explanatory and typical results are given in Table 5.7. The program is very quick to run and is well suited to a small personal computer. It should be noted that the crosspolar values will only be strictly valid for a practical horn with a large number of corrugations per wavelength and approximately equal slot width/ ridge width as discussed in Section 5.3.

Chapter 6

Manufacture and testing of corrugated horns

6.1 Manufacture of corrugated horns

Corrugated horns are not easy, nor cheap, to manufacture. Considerable manpower needs to be devoted to the production of a horn. Changes in the design after manufacture are hard to implement and there is an obvious desire to produce a design, which, when made and tested, will agree with the theoretical predictions. The corrugations are designed for operation over a specified frequency band. It thus immediately follows that all dimensions and tolerances on dimensions need to be considered in relation to wavelength rather than absolute distance. We will deal with tolerances in more detail later in this section. To start with we shall discuss some of the methods which have been used to manufacture corrugated horns.

Machining

Making the horn by machining, Fig. 6.1, is the most common method of manufacture for horns designed to operate at microwave frequencies, say below 20 GHz. Aluminium is the preferred material because of its relatively low density, but sometimes brass is used for small or intricate horns. Long horns need to be made in a number of sections and joined together. The part of the horn which presents the greatest difficulty in machining is the throat region. The requirement of an approximately $\lambda/2$ deep slot at the throat, when the inner diameter of the corrugated section is only about one wavelength diameter, limits the size and shape of the cutting bar which can be inserted into the horn. This is particularly a problem when the slot is narrow (i.e. a few millimetres). In this case the cutting tool tends to drag on the metal, either distorting the walls of the slot or giving an uneven surface to the slot walls and base. Since one mistake can ruin the complete horn, considerable care is needed when cutting the throat section. One solution is to make the horn in a number of separate sections and join them together. We have made horns successfully by this method, even when each section consisted of one slot and one ridge, Fig. 6.2, but there are disadvantages. The outer diameter needs to be increased in order to take the fixing screws. This increases the weight.

Each section needs to have ends which are very flat and perpendicular with the horn axis, otherwise some slight bending of the horn can occur. Since this will excite azimuthally dependent modes it must be avoided. Each section needs to be bedded firmly to the next section so that there is no gap anywhere around the periphery. A gap will lead to a discontinuity in the current flow which then distorts the fields in the slot and can cause effects out of all proportion to the gap size.

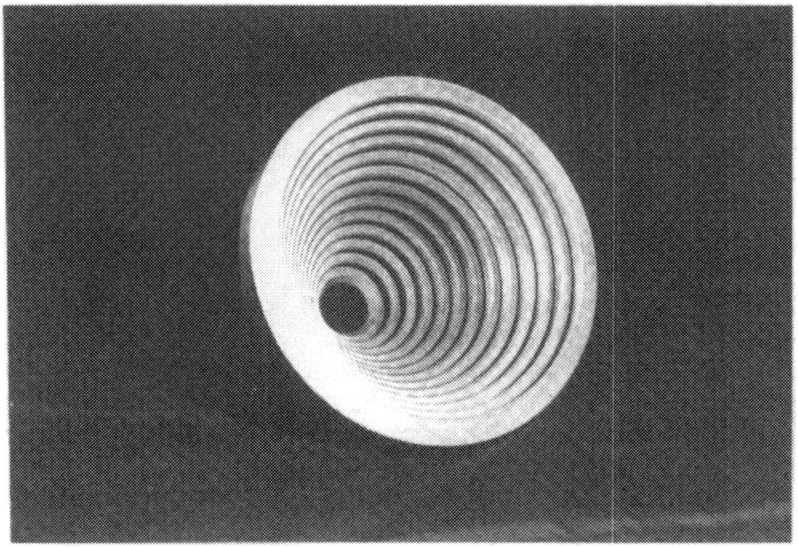

Fig. 6.1 *Corrugated horn machined in one piece*

We have used a thin conducting paste between sections to electrically seal the gap between sections.

Electroforming

Corrugated horns have been successfully electroformed, particularly for use at millimetre wavelengths. A mandrel of aluminium is turned with the inverse groove pattern and then copper, or other metal, is electroformed onto the mandrel. The mandrel is finally removed with a solvent. The problem is to ensure that the deposited metal is 'thrown' evenly down to the base of the slots in the mandrel. For this purpose it is better to have a horn with equal slot and ridge widths. At the throat region the slots cannot be very narrow, otherwise the metal will not go down to the base of the half wavelength deep slots.

A hybrid electroforming/machining technique has been described by Dragone [54] for making a corrugated horn, designed for 100 GHz. The mandrel is made up of brass discs (to form the ridges) and aluminium spacers (to form the slots). A copper wall is then electroformed onto the composite mandrel. It adheres to the brass discs but not to the aluminium. The inner is then machined to the correct cone and finally the aluminium spacers are removed by solvent. This method has the advantage that thin ridges can be made but the disadvantage that good contact around the base of the slots is difficult to guarantee. As mentioned above, interruption to the current flow can be highly damaging to the operation of the horn.

Fig. 6.2 *Corrugated horn manufactured in sections*

As the frequency of operation is increased above 100 GHz, the construction of mandrels require some care. Often slots are required which have widths of less than 200 μm. These are difficult to machine on a lathe. A spark erosion technique has been successfully used by Wilde*, Fig. 6.3. A computer controlled wire spark erosion machine is used to cut the slots. A thin brass or molybdenum wire is continuously moved over the surface of the slowly rotating mandrel. A high frequency pulsed voltage is applied to the wire causing controlled erosion of both the wire and the mandrel. The wire is continuously being replaced by fresh wire so it acts as a bandsaw. The process can cut slots 100 μm wide and 500 μm deep, making it feasible to construct corrugated horns at 300 GHz.

Other techniques

Small corrugated horns for use at a few gigahertz have been made by casting. Because the wavelength is relatively long at these frequencies it is usually unnecessary to do anything with the surface. The finish which can be obtained by casting is adequate for making a feed for a prime focus reflector intended for direct boradcast satellite reception. Casting is thus a cheap and simple way of making horns in large numbers.

* Wilde, R. (private communication)

Corrugated waveguides have also been made by a thermal contraction process. The ridges are formed from rings of brass or copper which are held on a specially constructed mount. The outer wall consists of a length of smooth wall circular waveguide. The brass discs have a slightly greater outer diameter than the inner diameter of the waveguide. The rings are cooled by immersion in liquid nitrogen so that they contract enough to allow insertion into the circular waveguide. When

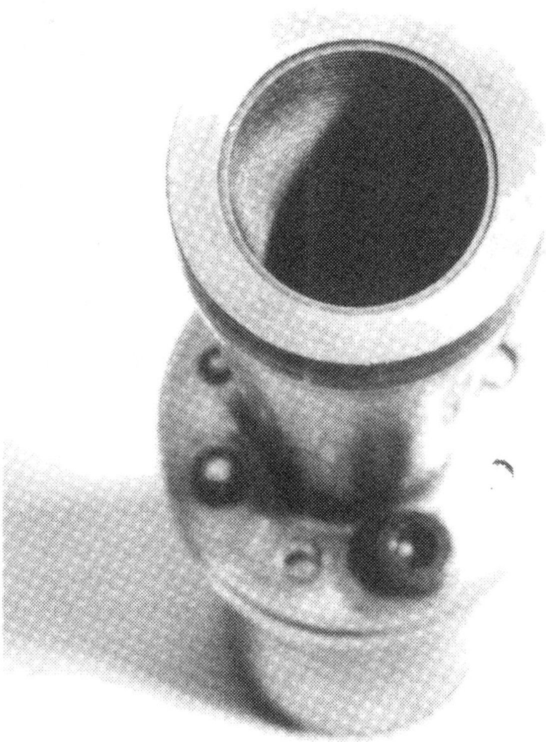

Fig. 6.3 *Millimetre wave corrugated horn made by electroforming for use at 100 GHz (Courtesy Thomas Keating Ltd.)*

the rings re-expand they grip the wall of the waveguide. The method was used to make lengths of waveguides for low attenuation experiments at 10 GHz. Brass rings (coefficient of expansion $15{\cdot}7 \times 10^{-6}/^{\circ}C$) of diameter 100·065 mm, were inserted in a copper waveguide of inner diameter 100 mm, Fig. 6.4. The samples worked well in the laboratory showing that the method does give good contact between ridges and wall.

An alternative to the above method is to use r.f. heating as a way of soldering rings (to form ridges) to an outer smooth wall.

Tolerances

The tolerances on a corrugated horn can be considered in three ways; the absolute tolerances in millimetres, the relative tolerances in wavelengths and the consistency of manufacture from one part of the horn to another part. It has already been mentioned that the absolute tolerances are not as important as the relative tolerances in wavelengths. Thus the absolute tolerances become progressively tighter as the operating frequency is increased. However, the relative tolerances do not need to be as tight as is often supposed. We will now examine briefly the effects of changes in each of the horn diameters.

Fig. 6.4 *Rings which will form ridges in corrugated waveguide made by thermal contraction*

Changes in the horn aperture will modify the copolar radiation patterns but hardly affect the crosspolar performance. Reference to the beamwidth curves, Figs. 5.10–5.12, shows that changes in diameter produce a proportional change in beamwidth. For corrugated horns, at the balanced hybrid frequency an approximation to the main beam copolar pattern is given by the function $\cos^7(0{\cdot}48\,(D/\lambda)\theta)$ where θ is in degrees. This allows an easy assessment to be made of changes in pattern shape. For instance, a 5% increase in the diameter of a two wavelength horn will decrease the -10 dB beamwidth by 5%.

Changes in the slot depth influence the crosspolar radiation characteristics as reference to Figs. 5.22–5.25 will show. The slot depth determines the resonant frequency at which minimum crosspolarisation occurs so changing the slot depth moves this minimum with frequency. The above figures can be used to assess the affect, which is quite severe for large diameter horns. However, in compensation, a large diameter horn has a wide crosspolar bandwidth, so the change in frequency of the minimum is less important than appears.

Changes in the slot width and ridge width have less affect than the slot depth, but again they move the crosspolar minima, as shown by Figs. 5.22–5.25. They have negligible influence on the copolar patterns over the whole range of feasible values.

The consistency of all the dimensions along the horn is far more important than the relative or absolute tolerances. An error in the dimensions of only one slot acts as a discontinuity in an otherwise smoothly changing horn. The discontinuity can be expected to generate high-order modes if it is at a section of the horn where the higher-order modes are capable of propagating. The performance of the horn is then adversely affected due to one error. We have experienced many cases, particularly at high frequencies, where the horn has been rendered useless by one discontinuity. It seems that horns made by electroforming are more likely to have isolated errors than machined horns, perhaps because the electroforming has failed to form one part of one slot.

Shape of corrugations

There are a few features concerning the shape of the slots which need discussing. It has already been mentioned that there is no theoretical reason why the corrugations should be cut perpendicular to the horn axis, Fig. 6.5(a), or perpendicular to the walls of the horn, Fig. 6.5(b). Since it is normally easier to make narrow flare angle horns with the former slots, this has become common practice.

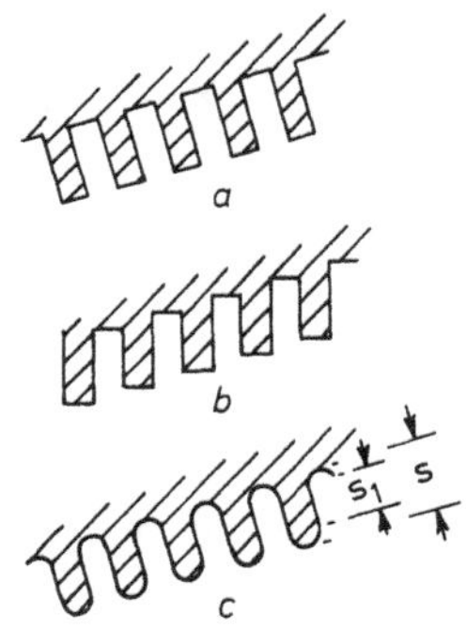

Fig. 6.5 *Shapes for corrugations*

The electrical depth of the slots is not always easy to assess, particularly for wide slots and wide flare angle horns. It seems best to use the mean depth of the slot at its centre as the design depth.

Although it has been assumed throughout this text that the slots are rectangular in shape, this does not have to be the case. The corrugations can be some other shape, as long as they are consistently the same throughout the complete length of the horn. Al-Hariri [4] investigated the effect of different slot shapes by modelling the slot with a staircase representation of rectangular shaped slots. He then used a radial transmission line approach to derive the equivalent admittance looking into the slots. The results for various slots shapes are shown in Fig. 6.6.

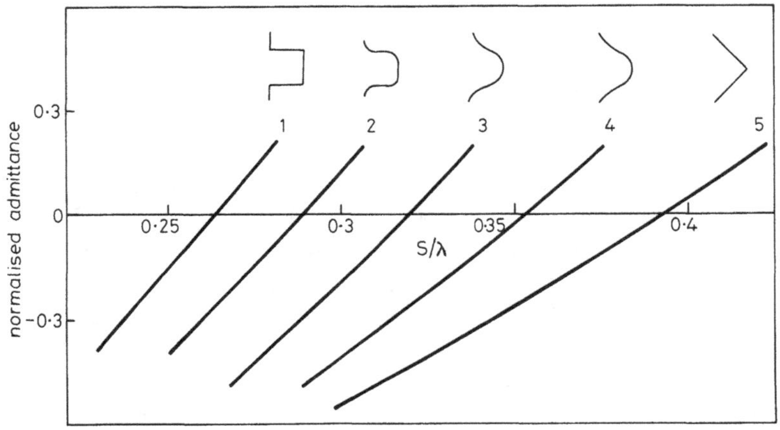

Fig. 6.6 *Normalised admittance against normalised slot depth for shaped slots*
1 Square 2 Arcs of circles plus flats 3 Arcs of circles 4 Cosine 5 Triangular

The assumption is made of a large number of slots per wavelength. This indicates that any of the shapes shown will give a zero normalised admittance at $r=r_1$ provided the depth of the slot is increased. Triangular slots need to be nearly 0·4 λ deep and arcs-of-circle slots (curve 3) 0·32 λ deep. This behaviour is expected from an intuitive assessment of the electrical characteristics. The equivalent inner diameter is also increased by the shaped slots. We have measured the radiation characteristics of a length of open-ended corrugated waveguide (1·7 λ diameter) with approximately cosine shaped slots. Apart from the shift in operating frequency to a higher value compared to the rectangular slot waveguide, the radiation characteristics were undistorted.

The effect of bevels and corners, Fig. 6.5(c), on the edges of the corrugations can also be assessed by the above technique. We have found that the effective slot depth does not change significantly as long as the ratio s_1/s is greater then 0·8. Again the bevels should be on all corrugations along the horn.

6.2 Testing of corrugated horns

6.2.1 Introduction

The radiation characteristics of corrugated horns need to be measured and compared to the theoretical predictions or the design specifications. In principle the testing of corrugated horns is no different from the testing of any other type of horn. However, the design specifications are often more stringent, particularly with regard to the crosspolarisation levels, than for other types of horn.

Most corrugated horns are used as feeds for reflectors, so that there are in principle two sets of measurements to be performed. Measurements on the feed horn and measurements on the reflector plus feed horn. However, it is not always practical to measure the characteristics of the reflector plus feed horn due to the large electrical size of the reflector and hence the large test range required for measurement, thus measurement of the feed horn alone is then the only measurement which can be undertaken. This is, in principle, adequate since modern test ranges and techniques enable the feed to be accurately measured and computer prediction can be used to give a good estimate of the total antenna radiation pattern.

The general procedure for testing antennas can be found in IEEE Standard Test Procedures for Antennas [184] and Appel-Hansen, Chapter 8 Rudge *et al.* [191]. The radiation characteristics of any test antenna are found by placing the antenna in a region of pseudo plane electromagnetic waves. The antenna is rotated and the power received by the antenna for specified orientations of the test antenna with respect to the plane waves are recorded, Fig. 6.7. This assumes that the test antenna is being used in a receive mode, the usual situation, but there is no reason why it should not be used as a transmitter if that is more convenient.

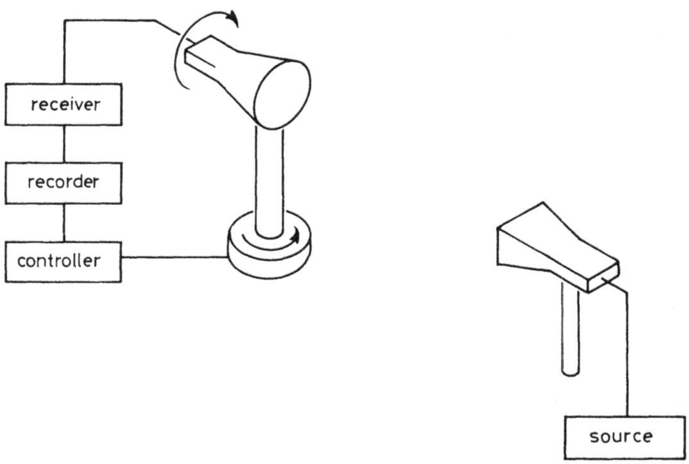

Fig. 6.7 *Antenna measurement system*

The recorded patterns will differ from the ideal pattern because the wavefront in the test region will not be truly 'plane', i.e. constant in amplitude and phase. The 'planeness' of the wavefront, for a given test antenna, is a measure of the performance of the test range. The nature of distortion away from a plane wave depends partly on the type of test range being used and partly on the quality of the test range. This latter factor is usually a function of the size and hence cost of the range. Because a test range can be very expensive to construct there is a natural desire to make a range which is only just adequate for the intended measurements. This is fine until more stringent specifications are required, in which case the performance of the range needs to be reassessed. Ideally an assessment of range performance would be carried out for each test antenna. This is impractical but the relationship of the test antenna to the test range should always be in the mind of the antenna engineer. Fortunately most corrugated horns are relatively small in size and do not require large test ranges.

6.2.2 Types of test range

There are, in principle, four types of test range which could be used for measuring horns:

outdoor far-field range;
indoor far-field range (anechoic chamber);
compact antenna range;
near-field or intermediate-field range.

The far-field range (outdoors or indoors) creates a pseudo plane wave region by utilising a small portion of a pherical wavefront, Fig. 6.8(a). If the test aperture diameter is small by comparison with the length of the range, the amplitude and phase of the electric field perpendicular to the axis of the range will taper only gradually from the axis. Usually the minimum length of a far-field range is taken as $2D^2/\lambda$ where D is the test aperture diameter, though for high quality horn testing, a length of, say $4D^2/\lambda$ is preferable.

Fig. 6.8 *Types of test range*
a Far-field range
b Compact antenna range

The outdoor far-field range is needed for testing large antennas. One, or both, of the source and test antennas are mounted on towers spaced some distance apart. It has many disadvantages. The performance of the range is difficult to evaluate, operation is susceptible to weather and other environmental factors, the land utilisation is high and it lacks security. However, for a large antenna there is usually no other viable alternative. Since most corrugated horns are relatively small in wavelengths, an outdoor range is neither necessary, nor recommended, for horn testing. The indoor far-field range or anechoic chamber is, by contrast, ideally suited for horn testing Measurement of the crosspolar radiation pattern requires a low inherent crosspolar level in the range and this can be achieved in an anechoic chamber. It also requires precise alignment of the test antenna with respect to the source antenna and this is more easily achieved in an indoor environment. The design of a suitable anechoic chamber is described below.

The Compact Antenna Range generates a pseudo plane wave region by collimating the radiation from a point source with a parabolic reflector, Fig. 6.8(b). A portion of the collimated beam will be a plane wave with superimposed small amplitude ripples caused by scattering from the edges of the reflector. Both high quality, high cost Compact ranges and medium quality, low cost Compact ranges, have been constructed and widely used. Antennas up to about 2 m in diameter can be measured in a compact, indoor environment. There is, however, a severe disadvantage as far as the measurement of corrugated horns is concerned. The offset paraboloidal reflector generates a crosspolar component and the Compact range reflector feeds have a relatively small aperture diameter. This means that their inherent crosspolarisation is high. These two factors mean that the range crosspolarisation is of the order of -30 dB, far too high to enable low crosspolar corrugated feeds to be measured. A dual offset reflector can offer a solution.

In near-field ranges the test antenna is usually the source antenna. The radiated field of the test antenna is sampled in its near field region by scanning a probe over a surface. The surface may be plane, cylindrical or spherical. The amplitude and phase of the radiated field is recorded, digitised and processed with algorithms to yield the far-field pattern. It is a very accurate method of antenna pattern measurement, particularly suited to cases where the antenna is mounted on bigger structures such as a satellite. It has not been widely used for feed horn testing, to date, but there is no reason in principle why the patterns of corrugated horns should not be measured by this technique.

6.2.3 Anechoic chamber design for low crosspolarisation horn measurement

An anechoic chamber should be designed to give a good plane wave region which will be large enough to cover the horn in all orientations and at all operating frequencies. It should have a level of background radiation, due to scattering off the walls of the chamber, which is at least 10 dB below the minimum pattern level to be measured. Thus, if the minimum copolar and crosspolar levels to be recorded are -40 dB then the chamber should have a background radiation level of -50 dB for its crosspolar scattering. The background radiation level in the chamber is measured by the *reflectivity* level, of which more later.

The main design factor is the length of the chamber which will determine the amplitude and phase taper across the test antenna. A range length which just satisfies the far-field criteria of $2D^2/\lambda$ gives a phase difference of $22\frac{1}{2}°$ between the centre and edge of the test aperture. The observed recorded effect on a radiation pattern with theoretically deep nulls is to fill in the nulls. For a uniformly illuminated circular test aperture the minimum value of the first null of the copolar pattern will be -27 dB instead of $-\infty$ dB and the first sidelobe level is increased by 0·2 dB. The amplitude taper is determined by the aperture size of the source horn. For a $2D^2/\lambda$ range, the angle subtended at the source by the test antenna is

$$\theta = \frac{D}{R} = \frac{\lambda}{2D} \tag{6.1}$$

With $D = 8\lambda$ this gives $\theta = 3{\cdot}6°$. Reference to the pattern of a horn (for instance, Fig. 5.4) shows that a 4λ diameter source horn will give an amplitude taper of 0·6 dB at the edge of the test aperture. This level of amplitude taper will increase the observed sidelobe level by about 0·2 dB and slightly increase the observed copolar beamwidth. A smaller source horn will generate a wider beam and hence a smaller amplitude taper, however, it will also increase the amount of energy incident on the walls of the chamber. Taken together the result of this discussion on the amplitude and phase taper is to indicate that a range length of greater than $2D^2/\lambda$ is preferable for high quality horn testing. Doubling the range length to $R = 4D^2/\lambda$ gives a test range which is capable of measuring deep nulls in radiation patterns and distorts the pattern shape by a negligible amount if a source horn of 4λ diameter is used.

The shape of the anechoic chamber can be of two main types, either a rectangular or funnel shape. The latter tapers from one end to reach a point at the other end. The amount of absorber needed on the walls is reduced by more than 50 per cent over the same length rectangular box. But its design has disadvantages. Most materials are relatively poor absorbers at high angles of incidence and reflect a proportion of the incident energy. The walls of the funnel chamber cause it to act similar to a pyramidal horn and consequently the wave propagating from the source at the apex is not a plane wave but a waveguide mode. This need not be detrimental as long as the phase centre of the horn source antenna is at the phase centre of the tapered chamber. Care in setting up and locating the source antenna is required.

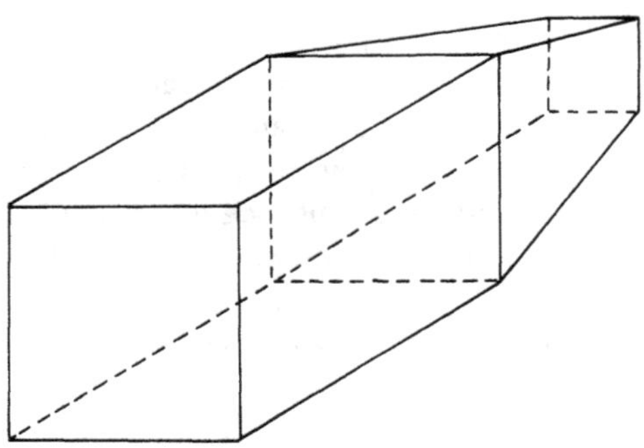

Fig. 6.9 *Sketch of Queen Mary College anechoic chamber*

The rectangular chamber has none of these technical problems. It is simple to operate and set up. In addition it can be used for scattering and near-field probing measurements.

A compromise between the two designs is possible which gives the advantages of the rectangular chamber and some of the cost saving of the tapered chamber. One half of the chamber, that used for the test region, can be rectangular in shape an the other half can be tapered to a smaller cross-section but not a point. We have built and successfully operated a chamber based on this compromise design. A sketch of the chamber is shown in Fig. 6.9. It is intended principally for use between 8 and 18 GHz. The chamber has a cross-section of 3 m by 3 m and tapers down to cross-section of 1·2 m by 1·2 m.

The cross-section at the quiet zone needs to be such that the test antenna is electrically 'isolated' from the walls. Energy scattered off the test horn and the mechanical support system should have a path loss of about 40 dB before reaching the walls. The cross-section of the chamber is made square in order to preserve symmetry. This is important for crosspolarisation measurements to ensure that most reflections have a similar path loss.

The walls of the anechoic chamber are completely covered with radar absorbing material. This should have a reflectivity at normal incidence of at least −40 dB for good crosspolar measurements. The absorber on the end wall has to absorb all the incident energy which is not recieved by the test horn. The absorber on this end wall should therefore have a better reflectivity than the side walls. In our chamber the end wall and the square box section are covered with −50 dB reflectivity absorber (at X band) while the tapered section is covered with −40 dB reflectivity absorber.

The horn used for the source can be any linearly polarised horn which satisfies the beamwidth criteria discussed above. It does not need to have a low crosspolarisation as long as the crosspolarisation is zero on boresight. Reference to any of the crosspolarisation patterns in the previous chapters will show that the crosspolarisation within the small angular region illuminating the test aperture is very low and much less than the level generated by reflection off the walls. For this reason a pyramidal horn can be used as the source antenna even though its peak crosspolar level is relatively high. A pyramidal horn generates a nearly perfect linear polarised field in the boresight direction. Pyramidal horns are readily available and so tend to be the preferred source horn for anechoic chambers.

The final component in the anechoic chamber is the turntable and mechanical support for the antennas. These need to be rigid for crosspolarisation measurements, but also have the smallest possible cross-section when viewed from the source antenna. Supports made from plastic or fibreglass can reduce the reflected signal. The complete pattern measurement of a corrugated horn requires two turntables; an azimuth turntable and a polarisation rotating turntable, Fig. 6.7. It is not necessary to have an elevation turntable but some form of elevation adjustment should be provided for alignment purposes. Similarly, a convenient means of adjustment in the vertical, horizontal and longitudinal directions will also aid alignment.

The correct alignment of the test horn with respect to the source antenna is one of the most time-consuming aspects of low crosspolarisation measurements. Copolar measurements are less critical and can usually be aligned by eye, but to be able to measure crosspolarisation patterns at levels below −30 dB requires precise alignment. The axis of the test horn must be exactly coincident with the axis of the source horn. Various optical instruments can be used but we have found that a visible light laser is a very convenient method of alignment. The laser is set up to shine through and along the axis of the source antenna waveguide and along the axis of the chamber. The source antenna is first aligned with the laser using temporary cross-wires as guides. Then the test antenna is carefully aligned by the same method.

The instrumentation required for the measurement of the copolar and crosspolar patterns is the same as any antenna pattern measurement, namely, stable source, stable reciever, pattern recorder and associated control instrumentation. The use of digital control and the immediate digitising of data for subsequent processing and display are increasing both the flexibility and the accuracy of the measurement sequence.

6.2.4 Assessment of test range performance

The basic performance of an anechoic chamber is determined by the energy scattered off the walls, floor and ceiling of the chamber. This can be understood by considering the situation shown in Fig. 6.10. The signal picked up by the test antenna will be the sum of two components, the direct signal and the signal reflected off the walls, i.e.

$$V = A_1 e^{j\alpha_1} + A_2 e^{j\alpha_2} \tag{6.2}$$

A_1 and A_2 are the amplitudes of the radiation pattern of the test antenna in the direct and reflected signal directions plus any path length loss differences. A_2 also includes the attenuation which the reflected wave has undergone at the walls of the chamber and normally $A_2 \ll A_1$. α_1 and α_2 are the path phase lengths of the respective directions plus any test pattern phase differences in the two directions.

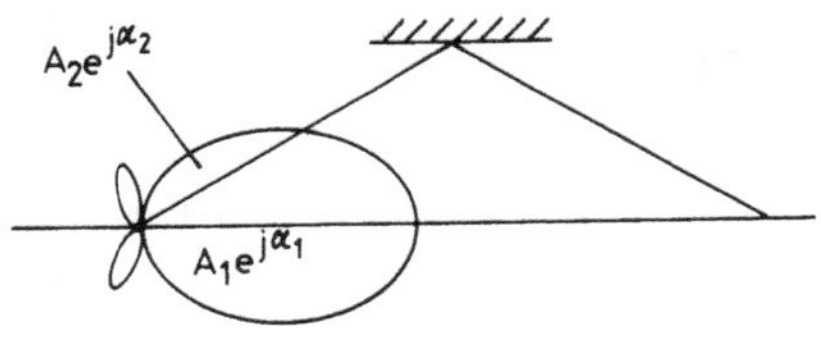

Fig. 6.10 *Spurious reflections from wall of anechoic chamber*

Now if the test antenna is moved sideways, the path length α_2 will change considerably compared to α_1. The signal received at the test antenna will oscillate between a maximum value ($|A_1| + |A_2|$) and a minimum value ($|A_1| - |A_2|$). The peak-to-peak deviations of the received signal will be:

$$t = 20 \log_{10} \frac{|A_1| + |A_2|}{|A_1| - |A_2|} \tag{6.3}$$

In practice there will be a number of unwanted reflected components. This will not affect the principle of the above and if the peak-to-peak deviation in the recieved signal is measured as the test antenna is moved around the test area, then the total peak-to-peak deviations recorded will be given by:

$$t = 20 \log_{10} \frac{|A_1| + |y|}{|A_1| - |y|} \tag{6.4}$$

The quantity y is the ***reflectivity*** level of the test range for direct signal level A. This equation is plotted in Fig. 6.11. Note that the reflectivity level, as here defined, is a function of a particular test antenna since the value is partly determined by the radiation pattern. A test antenna with a near omnidirectional pattern will receive unwanted components from more directions simultaneously than a highly directional antenna. It is therefore important to assess the chamber performance with a test antenna which has similar radiation characteristics to the horns to be measured.

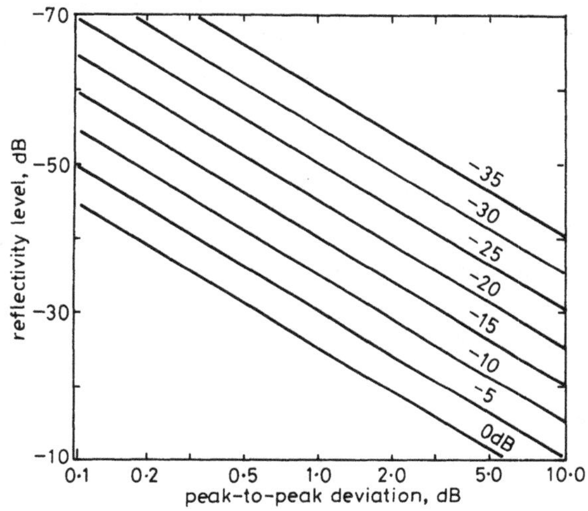

Fig. 6.11 *Reflectivity graph for use with pattern comparison assessment. Parameter: pattern level, dB*

The reflectivity level can be measured in two ways, by either the VSWR method or the Pattern Comparison method. Both methods are described in detail in Appel-Hansen (Chapter 8 of Reference 191). In the VSWR method, the field in the test region is recorded by moving a test antenna continuously across the test region. The test antenna is mounted on a movable carriage and the method is similar to sampling the VSWR in a transmission line using a moving probe. It is capable of giving a detailed picture of the test region field but requires specialised equipment which must be set up in the chamber. Also great care is needed to ensure that there are no extra reflections from the movable carriage which must necessarily be large to be rigid.

The pattern comparison assessment is easier to implement and requires only standard antenna test equipment. The test antenna is mounted on a turntable and its radiation pattern recorded. The turntable is physically moved to another position in the chamber and the pattern recorded, the positions are chosen to maximise the chance of observing maximum amplitude variations. All the recorded pattern are superimposed on top of one another, Fig. 6.12, and at various signal levels below the peak on a reference pattern, the maximum difference between patterns is noted. The signal level is $|A_1|$ and the maximum difference is t, hence eqn. 6.4 can be used to compute the reflectivity level y. A large number of values are measured and an average value deduced for the reflectivity level. For cross-polarisation measurements the procedure must be carried out twice, once for the copolar reflectivity level and then again for the crosspolar reflectivity level. The test antenna need not have low peak crosspolarisation but it should go to a low level at some azimuthal angles.

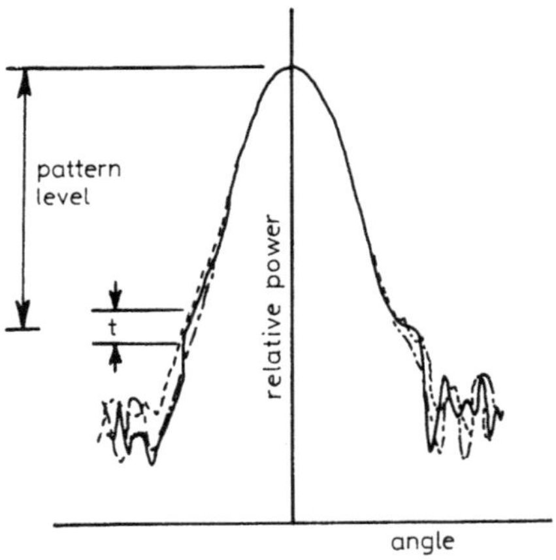

Fig. 6.12 *Superimposition of patterns*

Experience shows that the pattern comparison method tends to given a lower reflectivity level than the VSWR method, probably because the test region field is being sampled rather than continuously probed. However, it is a more convenient method to implement and the reflectivity level can be assessed using the actual corrugated horn for which accurate measured patterns are desired.

6.2.5 Measurement of copolar and crosspolar patterns

The co-ordinate system used for radiation pattern measurements is shown in Fig. 6.13. The radiation pattern is always recorded by rotating the test antenna in the $\phi = 0°$ plane (xz plane). The patterns in various ϕ planes are obtained by rotating the source and test antennas about the z-axis so as to align the electric field vector at the desired angle ϕ with respect to the xz plane. This scheme means that the boresight copolar signal level is constant for all planes. The measurement procedure for the amplitude patterns of a corrugated horn with a relatively low crosspolarisation is as follows:

(i) The corrugated horn to be measured is set up on the turntable and carefully aligned so that the axis of the test horn and the axis of the source horn are coincident. If the horn pattern has a crosspolar null in the boresight direction (as is normally the case), this fact can be used as an aid to alignment. The test horn should be placed such that its phase centre is over the centre of the turntable, then rotation of the test horn will keep the electrical path lengths from the source to the test horn constant at different azimuth angles. In practice, the centre is not uaually known precisely and so a guess has to be made. If only far-field patterns are required then the precise location of the test horn with respect to the turntable is not important. If phase patterns or near-field amplitude patterns are required then the electrical phase centre must be found, see below. Note that, if the phase centre of the horn is not at the aperture of the horn, then the effective test aperture, in the xz plane will be larger than the horn aperture.

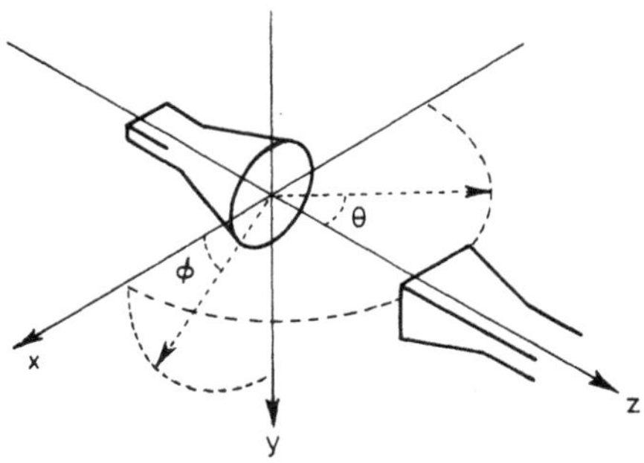

Fig. 6.13 *Co-ordinate system for feed measurement*

(ii) The H-plane copolar pattern is recorded by orientating the electric field vector of both the source and test antennas so that they are parallel with the y-axis. Then rotate the test antenna in the xz plane and record the pattern over the desired angular range of θ. The gain controls of the receiver should be set up to give an undistorted output with enough dynamic range for the measurements. The variable gain controls should not be adjusted after this initial set-up. All recorded signal levels will then be with respect to the peak copolar level. Any change in gain should be done using calibrated switched attenuators.

(iii) The H-plane crosspolar pattern is next recorded by rotating the test horn through 90° in ϕ so that its electric field vector is now parallel with the x-axis. The source antenna is unchanged. The test antenna is again rotated in the xz plane to record the crosspolar pattern. Of course this pattern level should be zero, so that only the inherent reflectivity level of the chamber will be recorded. In practice, it may be difficult to find the H plane copolar orientation in (ii) precisely, so use can be made of the knowledge of zero crosspolarisation in the H plane by reversing (ii) and (iii). The H plane crosspolar orientation is found by adjusting the test horn ϕ angle until no signal is received. This then defines the direction of the electric field vector in the test horn. The test horn can now be rotated through precisely 90° in ϕ to record the copolar pattern. This scheme implies that there is an accurate method of measuring the angle of rotation of the test horn about the z-axis, a necessity for good crosspolar measurements.

(iv) The E plane copolar pattern is recorded by orientating the electric field vector of both source and test horns so that they are parallel with the x-axis, and again rotating the test horn in the xz plane.

(v) The E plane crosspolar pattern (which should be zero for single mode horns with unity azimuthal dependence) is found similarly to (iii) above.

(vi) The 45° plane copolar pattern is recorded by orientating the electric field vector of both source and test horn so that they are parallel to the $\phi = 45°$ or 135° direction. The pattern is again recorded by rotating the test horn in the xz plane. This case is shown in Fig. 6.14(a).

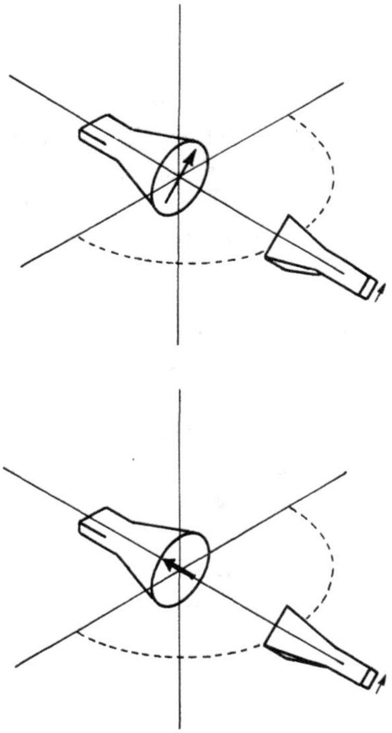

Fig. 6.14 *Test and source orientation for 45 degree plane measurements*
a Copolar
b Crosspolar

(vii) The 45° crosspolar pattern is recorded by turning the test horn through precisely 90° in ϕ, leaving the source horn unchanged, and repeating the θ pattern rotation. The situation is shown in Fig. 6.14(b). This measurement is one of the most important recordings for single mode horns with unity azimuthal dependence because the 45° crosspolar pattern should give the peak crosspolar level. The pattern levels recorded are with respect to the copolar peak level.

(viii) The patterns in any other plane can be similarly recorded by repeating the above procedures. A complete contour plot requires that the patterns are measured in enough planes to give a smooth record of the horn radiation characteristics.

6.2.6 *Measurement of phase patterns and gain*

The phase patterns need some form of phase-amplitude receiver. Usually the purpose of the phase measurements is to locate the phase centre of the horn and to see how nearly spherical is the copolar main beam. High sensitivity is not therefore needed, and a network analyser can be used. The shape of the recorded phase patterns will depend on the location of the test horn phase centre with respect to the axis of the azimuth turntable. Thus it is first necessary to locate the phase centre, as discussed in Section 5.4. The phase centre changes with azimuth angle θ, so it is generally assumed that the horn phase centre means the phase centre along the boresight direction. If the crosspolarisation is significant, the phase centre may be different in the E and H planes.

The phase centre can be located by recording the phase pattern over a narrow angular range about the boresight direction, then either move the horn along the z-axis and repeat the phase pattern measurement until the phase is constant, or use the recorded pattern and geometry to locate the phase centre.

After location of the phase centre in a particular plane the phase pattern in that plane can be recorded, indicating the way in which the horn radiation pattern deviates from a spherical wave.

The gain of a corrugated horn can be measured by any of the standard methods described in IEEE Standard Test Procedures for Antennas [184], and Rudge *et al.* [191]. The gain transfer method where the gain of the test horn is compared to the gain of a standard horn is easy to implement in an anechoic chamber. The gain of the standard horn should be similar to the gain of the test horn in order that the extraneous signals in the chamber affect both horns similarly. The gain needs to be known at all specified frequencies, otherwise another method such as the three antenna method must be used. In this case three horns are measured in all combinations and the free-space path loss formulae used to compute the gain of the test horn.

Chapter 7

Rectangular and elliptical corrugated horns

7.1 Introduction

Corrugated horns and feeds with non-circular cross-sections have received relatively little attention in the past. Out of nearly 180 papers cited in the bibliography only a few have dealt with horns of rectangular or elliptical cross-section. There are good reasons for this bias, for it turns out that unless a shaped beam is required, circular corrugated horns are superior in almost all respects to non-circular corrugated horns. Furthermore, rectangular and elliptical corrugated horns are difficult to analyse. The simple theoretical models are only an approximation to the true electromagnetic behaviour of the horns and do not give good predictions of cross-polar performance. The electromagnetic fields in the horns are more complicated and the fields in the slots are dispersive. The throat region is difficult to design so as to suppress unwanted mode excitation. Finally, the horns are difficult to manufacture and thereby expensive to produce. But, as stated, their one significant advantage compared to circular corrugated horns in certain applications, is their ability to generate a beam with a non-circular symmetry. This is desired in some spacecraft antennas and also some radar antennas. In this chapter we shall discuss their properties but the approach will be limited to a study of the basic characteristics, without the detail which has been devoted to circular corrugated horns.

7.2 Rectangular corrugated horns

7.2.1 Background

The first paper on rectangular corrugated horns appears to be that of Lawrie and Peters [92] who introduced corrugations onto the E plane walls as a way of tapering the aperture electric field to reduce the E plane sidelobes. The two-walled rectangular corrugated horn can be analysed exactly, and a theory has been given by Baldwin and McInnes [11]. A number of authors have subsequently reported applications of the two-wall corrugated horn [50, 96, 113, 135, 142]. The first analysis of the four-walled rectangular corrugated horn with corrugations on all four walls was presented by Bryant [20] who used it as a horn for producing circularly polarised patterns with low sidelobes [19]. The four-walled corrugated horn cannot be analysed exactly using model analysis because the boundary conditions cannot be made coincident with the reference direction of a rectangular co-ordinate system. Bryant [20] used a superposition technique and treated the four-walled waveguide as the sum of two two-walled corrugated waveguides, an approximate procedure which ignores the influence of the corners of the slots. The theory presented by Bryant was not rigorous and started a controversy. Dybdal *et al.* [57] claimed that it was incorrect and subsequently [58] published an impedance compatibility relationship. This showed that the four-walled corrugated horn did not satisfy the relationship and so they deduced that it could not carry any power. However, they neglected the evanescent modes in the slots and the theory was shown by Isaac [73], Narasimhan [112] and Bach Anderson [8] to be incorrect. Since then no major advances in the theory have been reported and the best method of modal analysis is still that based on Bryant's superposition method with an additional orthogonal mode set.

A number of papers have reported applications of rectangular corrugated horns. The ability to shape the radiation pattern with a four-walled corrugated horn having the corners filled in was reported by Manwarren and Farrar [96]. Its use as a feed horn was reported by Baldwin and McInnes [14] and Davis [50]. The attenuation properties of two-walled corrugated waveguides were studied by Al-Hariri *et al.* [3, 4] and Baldwin and McInnes [10]. A four-walled corrugated horn was built and extensively measured in order to investigate its copolar and crosspolar radiation characteristics by Adatia *et al.* [2]. They found that it was necessary to include an EH set of modes, in addition to the HE modes to account for the crosspolar levels, and that higher order modes deteriorated the performance. The throat region of the horn was particularly difficult to design correctly, partly due to the overmoded nature of the waveguide and partly due to the lack of a reliable theory.

In the following analysis we shall start by studying the propagation characteristics of two-wall rectangular corrugated waveguides. The four-walled corrugated waveguide and horn can then be analysed by superimposing an orthogonal mode set. The corrugated walls are treated as an anisotropic surface impedance. The impedance boundary conditions lead to hybrid modes in the waveguide, as in the circular corrugated waveguide. The hybrid modes have fields with only one of the transverse electric or magnetic components zero. The two mode sets use the x direction as the reference and have either $H_x = 0$ or $E_x = 0$. The former gives HE modes (or TE_x modes) and the latter EH modes (or TM_x modes). The TM_x/TE_x designation has been used by most workers but because it leads to confusion with the smooth-wall rectangular waveguide where the mode designation is referred to the z direction of propagation, we will refer to the modes as HE_{nm} and EH_{nm}. The subscript n refers to the number of variations in the x direction and the subscript m refers to the numerical order of modes, of type n, as frequency increases from zero. Modes of order n can have transverse electric fields which are either symmetrical or asymmetrical with respect to central axis. This leads to four sets of field equations (i.e.

symmetrical x, symmetrical y; asymmetrical x, asymmetrical y). Consequently for each mode set (HE or EH) there are two characteristic equations and two sets of cut-off frequencies. The mode structure is thus considerably more complicated than circular waveguides, though because the fields are represented by trigonometric functions rather than Bessel function the characteristic equations are quicker to compute.

7.2.2 Two-walled rectangular corrugated waveguide

The analysis of the two-wall corrugated waveguide shows most of the features of the more general four-wall waveguide so will be studied here. A detailed analysis would require that the fields in the central region are represented by a fundamental wave and a set of space harmonics and the fields in the slots by an infinite sum of standing waves, most of which are evanescent. However, we shall assume that the space harmonics are zero and that fields in the slots are due only to the lowest order TE_{10} mode. It will be assumed that there are a large number of slots per wavelength and that the ridges are infinitely thin so that the longitudinal geometry is not a parameter in the analysis. The relevant geometry of the waveguide is shown in Fig. 7.1. At the surface of the corrugations, $y=\pm b/2, E_x=0$ and $E_y \neq 0$. Thus the boundary conditions will be satisfied by modes which are TE to x or HE modes.

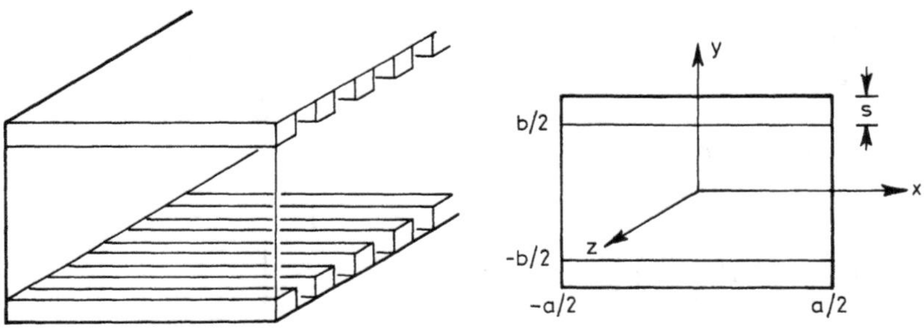

Fig. 7.1 *Two-walled rectangular corrugated waveguide*

The waveguide may be conveniently analysed by starting with a Hertzian magnetic vector potential (time dependence $e^{j\omega t}$ assumed).

$$F_H = \psi_H(x, y)\, e^{-j\beta z} \mathbf{a}_x \tag{7.1}$$

$$\mathbf{E} = -j\omega\mu\nabla\cdot F_H \tag{7.2}$$

$$\mathbf{H} = k^2 F_H + \nabla(\nabla\cdot F_H) \tag{7.3}$$

$$\nabla_T^2\psi_H + K_c^2\psi_H = 0 \tag{7.4}$$

where $k = 2\pi/\lambda_0$ and $K_c^2 = k^2 - \beta^2$. The solution to the wave eqn. 7.4 is

$$\psi_H = [A\cos(K_x x) + B\sin(K_x x)]\,[C\cos(K_y y) + D\sin(K_y y)]\, e^{-j\beta z} \tag{7.5}$$

where $K_c^2 = K_x^2 + K_y^2$.

Applying the boundary conditions that $\partial\psi_H/\partial_x = 0$ at $x = \pm a/2$, leads to $K_x = n\pi/a$ and four possible solutions:

$$\psi_{H_1} = A_n \cos\left(\frac{n\pi}{a}x\right)\cos(K_y y) \qquad n,\text{odd} \tag{7.6}$$

$$\psi_{H_2} = B_n \cos\left(\frac{n\pi}{a}x\right)\sin(K_y y) \qquad n,\text{odd} \tag{7.7}$$

$$\psi_{H_3} = C_n \sin\left(\frac{n\pi}{a}x\right)\cos(K_y y) \qquad n,\text{even} \tag{7.8}$$

$$\psi_{H_4} = D_n \sin\left(\frac{n\pi}{a}x\right)\sin(K_y y) \qquad n,\text{even} \tag{7.9}$$

ψ_{H_1}, ψ_{H_3} are symmetric in y and ψ_{H_2}, ψ_{H_4} are asymmetric in y. ψ_{H_1}, ψ_{H_2} are symmetric in x and ψ_{H_3}, ψ_{H_4} are asymmetric in y. The field components are obtained from eqns. 7.2 and 7.3. For ψ_{H_1} (symmetric, symmetric) these are:

$$\left.\begin{aligned}
E_x &= 0\\
E_y &= -\omega\mu\beta A_n \cos\left(\frac{n\pi}{a}x\right)\cos(K_y y)\\
E_z &= -j\omega\mu K_y A_n \cos\left(\frac{n\pi}{a}x\right)\sin(K_y y)\\
H_x &= \beta_1^2 A_n \cos\left(\frac{n\pi}{a}x\right)\cos(K_y y)\\
H_y &= \frac{n\pi}{a}K_y A_n \sin\left(\frac{n\pi}{a}x\right)\sin(K_y y)\\
H_z &= j\frac{n\pi}{a}\beta A_n \sin\left(\frac{n\pi}{a}x\right)\cos(K_y y)
\end{aligned}\right\} \tag{7.10}$$

where $\beta_1^2 = k^2 - \left(\frac{n\pi}{a}\right)^2$. Hence $K_y^2 = \beta_1^2 - \beta^2$

The impedance parallel to the surface of the corrugations at $y = \pm b/2$ is

$$Z = \frac{E_z}{H_x} = \frac{-j\omega\mu K_y}{\beta_1^2}\tan\left(\frac{K_y b}{2}\right) \tag{7.11}$$

In the slots we assume a cut-off TE_{10} mode, so $K_y = \beta_1$ and from eqn. 7.10 at

$y = \pm b/2$:

$$E_x = E_y = H_z = 0$$

$$E_z = -j\omega\mu E_1 \cos\left(\frac{\pi}{a}x\right) \sin(\beta_1 s)$$

$$H_x = -\beta_1 E_1 \cos\left(\frac{\pi}{a}x\right) \cos(\beta_1 s) \qquad (7.12)$$

$$H_y = \frac{n\pi}{a} E_1 \sin\left(\frac{\pi}{a}x\right) \sin(\beta_1 s)$$

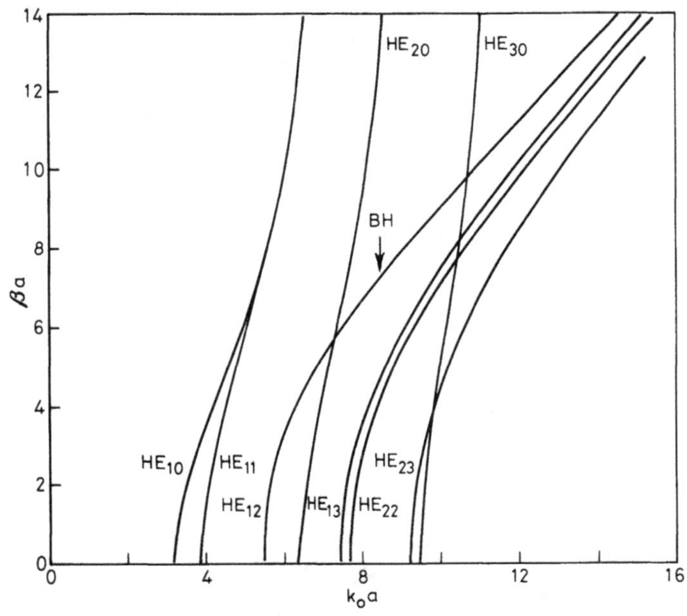

Fig. 7.2 *Dispersion curves for HE modes in rectangular corrugated waveguide: b/a = 1, s/a = 0·2, BH = 'balanced hybrid' frequency*

The surface impedance is

$$Z_s = \frac{j\omega\mu}{\beta_1} \tan(\beta_1 s) \qquad (7.13)$$

Note that Z_s is a function of s, a and frequency. It is therefore dispersive, in contrast to circular corrugated waveguides where the surface impedance is non-dispersive. The purpose of the corrugations is to create a 'magnetic wall' at $y = \pm b/2$, i.e. $H_x = 0$ or $\beta_1 s = m + \pi/2$ or $s/\lambda = 0{\cdot}25\sqrt{\{1 + 4(s/a)^2\}}$. The slot depth required to create this 'balanced hybrid' condition is thus dependent on the size of the main waveguide.

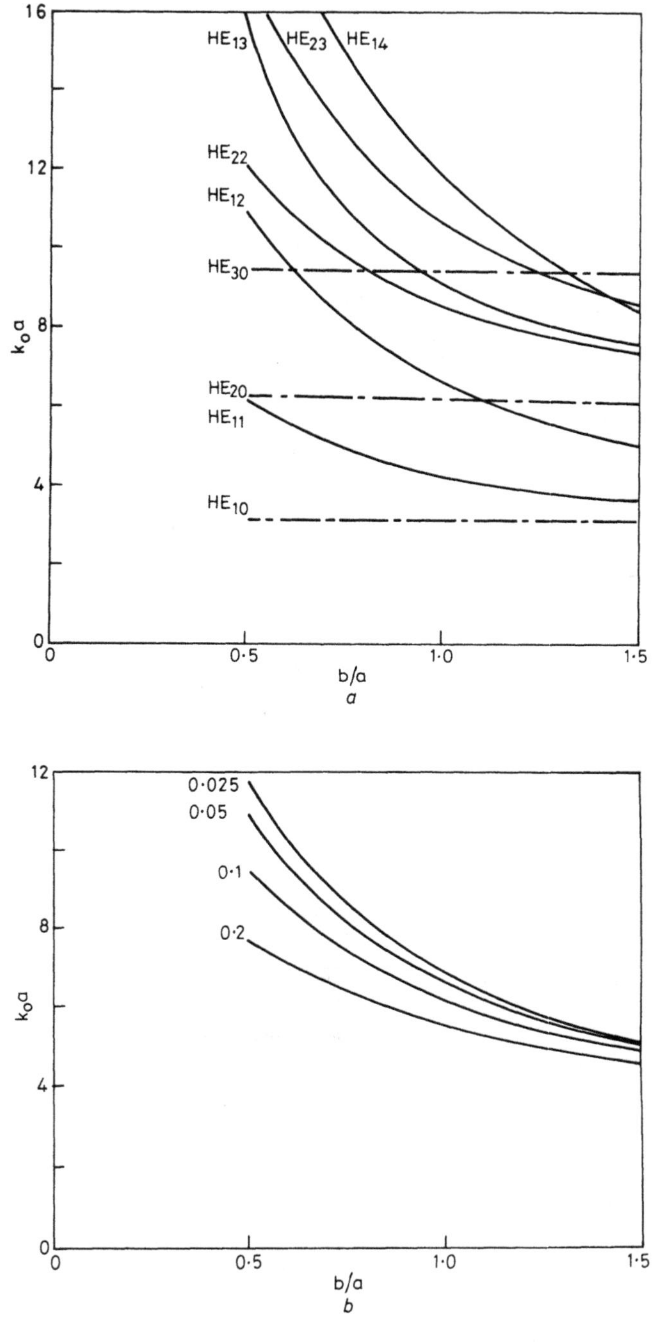

Fig. 7.3 *Cut-off characteristics for HE modes in rectangular corrugated waveguide*
a s/a = 0·05
b HE_{12} mode, parameter *s/a*

Equating eqns. 7.11 and 7.13 gives the characteristic equation

$$K_y \tan\left(\frac{K_y b}{2}\right) = -\beta_1 \tan(\beta_1 s) \tag{7.14}$$

This applies for all n and ψ symmetric in y. For ψ asymmetric in y, the characteristic equation is

$$K_y \cot\left(\frac{K_y b}{2}\right) = +\beta_1 \tan(\beta_1 s) \tag{7.15}$$

If β is greater than β_1, K_y becomes imaginary and eqn. 7.14 becomes

$$K_y \tanh \lambda\left(\frac{|K_y| b}{2}\right) = +\beta_1 \tan(\beta_1 s) \tag{7.16}$$

This equation gives slow waves bound to the corrugations. For computational purposes the equations are conveniently normalised to give for eqn. 7.14

$$(K_y a) \tan\left(\frac{K_y b}{2}\right) = (\beta_1 a) \tan(\beta_1 s) \tag{7.17}$$

Where $(K_y a)^2 = (\beta_1 a)^2 - (\beta a)^2$ and $(\beta_1 a)^2 = (ka)^2 - (n\pi)^2$. Also $K_y b = (K_y a)$ b/a and $\beta_1 s = (\beta_1 a) s/a$. This shows that the characteristics can be normalised to a and that the controlling geometric parameters are b/a and s/a. In practice the slot depth will be determined by the balanced-hybrid condition and consideration of the usual aperture sizes for horns shows that as a general guideline $0{\cdot}04 < s/a < 0{\cdot}2$.

Propagation characteristics for $b/a = 1{\cdot}0$ and $s/a = 0{\cdot}2$ are shown in Fig. 7.2. As stated earlier, the mode nomenclature is such that HE_{1m} modes are numbered consecutively from the lowest frequency. This is convenient but does mean that the field dependence is not immediately clear. Thus the HE_{10} mode is the symmetric slow wave solution of eqn. 7.16. The HE_{11} mode is the lowest order asymmetric solution of eqn. 7.15. The HE_{12} mode is the lowest order solution of eqn. 7.14. This mode is symmetric in both x and y and has a single cosine distribution in both planes. It is the mode with the first balanced-hybrid condition and is the mode required for low crosspolar radiation characteristics. The designation of the other modes is similar. There is no HE_{21} mode, corresponding to the HE_{11} solution.

A number of features of the HE_{12} mode are evident from Fig. 7.2 It is not the dominant mode in the waveguide and its solution at cut-off corresponds to a TE_{12}/TM_{12} mode in a smooth-wall rectangular waveguide. This implies that excitation of the mode in the throat region of a horn will not be as easy as in a circular corrugated horn. Also the lower frequency of operation should be chosen so that the HE_{10}/HE_{11} modes have gone to their high frequency cut-off. The region of operation is overmoded. This is for a small waveguide size. As the size becomes larger the overmoding increases considerably, and thus the possibility of higher order mode excitation becomes more likely.

The special points on the propagation characteristics can be identified by substitution into the characteristic equations. The cut-off frequencies ($\beta = 0$) are given for modes symmetric in y by:

$$(ka)^2 = \frac{p^2\pi^2}{\left(\frac{s}{a} + \frac{b}{2a}\right)^2} + (n\pi)^2 \tag{7.18}$$

and for modes asymmetric in y by:

$$(ka)^2 = \frac{(q+\frac{1}{2})^2\pi^2}{\left(\frac{s}{a} + \frac{b}{2a}\right)} + (n\pi)^2 \tag{7.19}$$

where $p = 0, 1, 2, \ldots$, and $q = 1, 2, \ldots$.

These are plotted in Fig. 7.3(a) and (b) for typical parameters. The higher frequency cut-off ($\beta \to \infty$) for all modes is given by:

$$(ka)^2 = \frac{(m+\frac{1}{2})^2\pi^2}{(s/a)^2} + (n\pi)^2 \tag{7.20}$$

The balanced hybrid frequency occurs when $\beta_1 s = \pi/2$ for the HE_{12} mode. This leads to the same condition as $\beta \to \infty$ and hence eqn. 7.20 also gives the balanced-hybrid frequency. It is a function only of s/a so as the waveguide size increases (implying smaller s/a) the balanced-hybrid frequency becomes correspondingly larger.

One potentially useful feature of the rectangular corrugated horn is that the balanced-hybrid conditions are maintained over a wide frequency range. This is shown in Fig. 7.4 where the wavenumber $K_y b$ is plotted against normalised frequency for the HE_{12} mode. The balanced-hybrid condition occurs when $K_y b = \pi$. It is seen that $K_y b$ is nearly constant over an octave of bandwidth for $s/a = 0{\cdot}1$. For a smaller s/a the bandwidth is even greater. However, the need to avoid the propagating band of the HE_{11} mode means that the useful frequency range is somewhat reduced.

7.2.3 Four-wall rectangular corrugated waveguide

The rectangular corrugated waveguide, shown in Fig. 7.5, with corrugations on all four walls is not amenable to exact analysis. The method used to obtain an approximate solution is to treat the waveguide as the superposition of two two-wall corrugated waveguides. With the principal electric field vector polarised along the y, or vertical axes, the electric fields of the HE type modes will see the vertical corrugations (containing a large number of corrugations per wavelength) as plane electric waveguide walls. Thus one of the two-wall corrugated waveguides is identical with the waveguide studied in the last section and all the results given there apply to HE modes in four-wall corrugated waveguides. The other two-wall corrugated waveguide has its corrugations on the side walls. Examination of the fields of the EH mode set to be supported by this waveguide, indicates that the top and bottom walls, at $y = \pm b/2$, should be treated as perfect magnetic conductors, while the impedance of the side walls at $x = \pm a/2$ will be due to the presence of a TE_{10} mode in the corrugations.

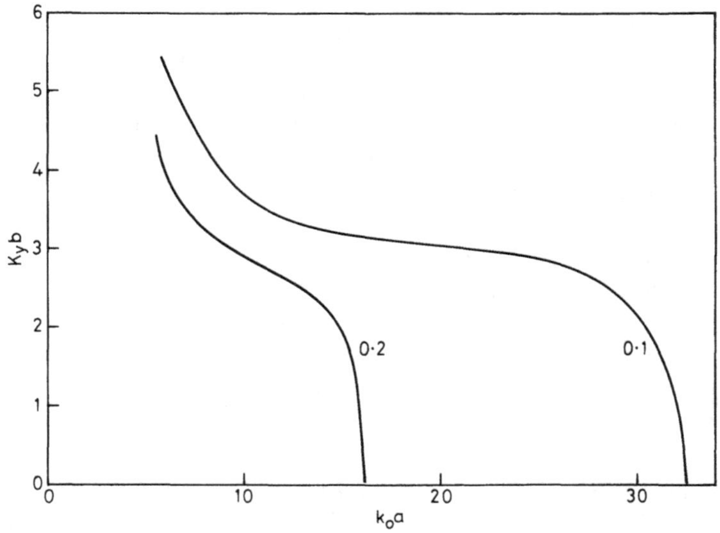

Fig. 7.4 *Wavenumber $K_y b$ against ka. Parameter s/a*

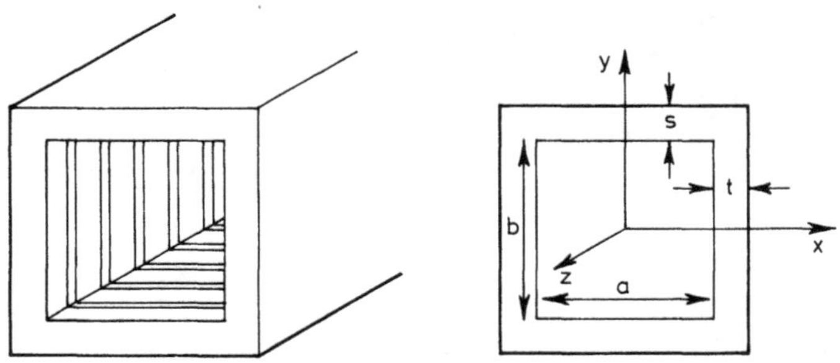

Fig. 7.5 *Four-walled rectangular corrugated waveguide*

The analysis is similar to the previous section, except that the boundary condition for eqn. 7.5 is $\psi_H = 0$ at $y = \pm b/2$. This leads to an impedance at $x = \pm a/2$ for modes symmetric in x of:

$$Z = \frac{E_z}{H_y} = \frac{j\omega\mu K_x}{\beta_2^2} \cot\left(K_x \frac{a}{2}\right) \tag{7.21}$$

where $\beta_2^2 = k^2 - (n\pi/b)^2$. The surface impedance due to the slots is:

$$Z_s = \frac{j\omega\mu}{\beta_2} \tan(\beta_2 t) \tag{7.22}$$

The characteristic equation follows as:

$$K_x \cot\left(\frac{K_x a}{2}\right) = +\beta_2 \tan(\beta_2 t) \tag{7.23}$$

Comparison with the characteristic equations of the HE modes shows that this has the same form as HE modes which are asymmetric in y, eqn. 7.15. EH modes with fields symmetric in x have a characteristic equation similar to eqn. 7.14. Thus in a square corrugated waveguide with constant slot depth on all walls the HE and EH modes will be degenerate. The propagation characteristics shown in Fig. 7.2 apply to EH modes with appropriate change of parameters.

The degeneracy of the HE and EH modes in a square corrugated waveguide is a serious disadvantage. Adatia *et al.* [2] have found that the EH modes are easily excited and contribute to the crosspolar radiation from a horn. Fortunately there is little reason to choose to use a square corrugated waveguide since a circular corrugated waveguide outperforms it in almost all respects. The main reason for considering a rectangular corrugated waveguide is to obtain different beamwidths with the electric vector polarised along the x and y axes. For $a \neq b$, the degeneracy disappears and the main practical effect of the EH modes is to increased the number of modes which can potentially propagate in the waveguide. Since the field patterns of the EH modes are similar to the HE modes there is a strong possibility of the modes coupling together.

7.2.4 Radiation characteristics of rectangular corrugated horns

The radiation patterns of a rectangular corrugated horn with narrow flare angle can be calculated using the Fourier Transform, or E-field, technique [132]. The far-field is:

$$F(x,y) = \iint (E_y \mathbf{a}_x - E_x \mathbf{a}_y) \exp\left[jk(x \sin\theta \cos\phi + y \sin\theta \sin\phi)\right] dx\,dy \tag{7.24}$$

where E_x and E_y are the aperture electric fields.

For the HE modes $E_x = 0$ and E_y is given by eqn. (7.10). Substituting into eqn. 7.24 and evaluating the integrals leads to the following expression for HE modes which are symmetric in y.

$$F_x = -2\omega\mu\beta A_n \left\{ \left[\frac{\dfrac{n\pi}{a} \sin\left(\dfrac{n\pi}{2}\right) \cos\left(\dfrac{ka}{2} \sin\theta\cos\phi\right)}{\left(\dfrac{n\pi}{a}\right)^2 - (k\sin\theta\cos\phi)^2} \right] \times \right.$$

$$\left[\frac{K_y \sin\left(\frac{k_y b}{2}\right) \cos\left(\frac{kb}{2} \sin\theta \sin\phi\right) - k \sin\theta \sin\phi \cos\left(\frac{k_y b}{2}\right) \cos\left(\frac{kb}{2} \sin\theta \sin\phi\right)}{[K_y^2 - (k \sin\theta \sin\phi)^2]}\right] \tag{7.25}$$

The copolar and crosspolar components of the far-field are given by

$$E_{co} = \left[\cos^2\left(\frac{\theta}{2}\right) - \sin^2\left(\frac{\theta}{2}\right) \cos 2\phi\right] F_x$$

$$E_{cr} = \sin^2\left(\frac{\theta}{2}\right) \sin 2\phi F_x \tag{7.26}$$

The far-field expressions for the asymmetric HE modes and EH modes can be similarly derived.

Since we are mainly interested in the behaviour around the balanced hybrid frequency of the HE_{12} mode, where $K_y b = \pi$, eqn. 7.25 can be simplified to:

$$F_x \propto (kb)(\beta a) \left[\frac{\cos\left(\frac{ka}{2} \sin\theta \cos\phi\right)}{\pi^2 - (ka \sin\theta \cos\phi)^2}\right] \left[\frac{\cos\left(\frac{kb}{2} \sin\theta \sin\phi\right)}{\pi^2 - (kb \sin\theta \sin\phi)^2}\right] \tag{7.27}$$

Examination of this equation shows that the first square bracket controls the H plane radiation pattern and the second square bracket controls the E plane radiation pattern. The two planes have identical forms so that a square horn will have a symmetric radiation pattern (ignoring the $\cos\theta$ factor in eqn. 7.26). This, of course, was the aim of the corrugations on the top and bottom walls. In a plain wall waveguide the E plane pattern characteristic is different and gives rise to an asymmetric pattern with very high sidelobes, as sketched on Fig. 7.6. This figure shows a universal pattern for the HE_{12} mode in a corrugated waveguide of side a operating near the balanced hybrid frequency. The pattern is relatively broad with a first sidelobe level of -24 dB. As expected, the efficiency of radiation from the physical aperture is low. Rectangular corrugated horns are bulky and possess a wide flange to encompass the corrugations which are in general much deeper than in their circular counterpart. Fig. 7.6 can be used to assess the patterns of a non-square horn by appropriate scaling of the factor a. From eqn. 7.27, the -3 dB half beamwidth in either the E or H planes is:

$$ka \sin\theta = 3{\cdot}74 \tag{7.28}$$

Thus the sine of the beamwidth is inversely proportional to the normalised height or the normalised width. This equation can be used as a design aid to obtain the required aperture dimensions. The equivalent equation for the -10 dB half beamwidth is $ka \sin\theta = 6{\cdot}4$.

Crosspolar results have not been computed using eqn. 7.27 because it is known [2] that the theoretical predictions considerably underestimate the level of cross-polarisation present in a practical horn. This is believed to be due to the presence of the EH modes. The accurate prediction of the crosspolar characteristics awaits a more sophisticated theory. Experimental models have given relatively high levels of crosspolarisation when compared to equivalent size circular horns, though some of this high level was due to higher order mode excitation. Note that the rectangular horn does not have a crosspolar characteristic with a bandpass type of frequency characteristic. This is in contrast to the circular corrugated horn where the crosspolar level theoretically goes to zero at one combination of radius, frequency and slot depth. In the rectangular horn case the crosspolar level decreases monotonically with frequency.

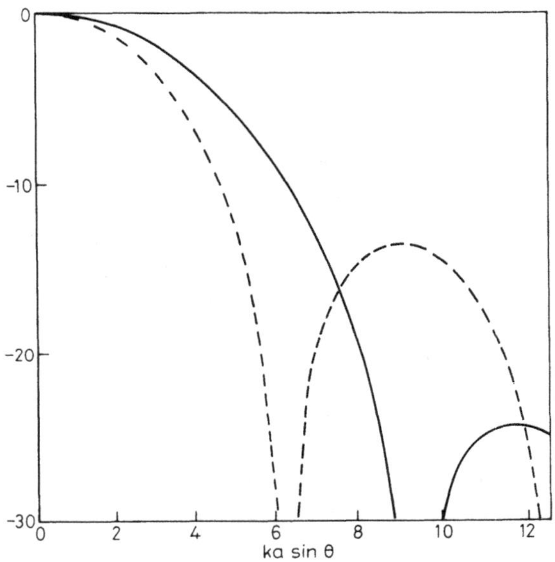

Fig. 7.6 *Radiation patterns for HE_{12} mode in rectangular corrugated waveguide*
——— H plane and E plane (ignoring cos θ factor)
– – – – E plane in smooth-wall rectangular waveguide

The above discussion has been confined to open-ended rectangular corrugated waveguides. The effect of the flare which is present on a horn is similar to the circular case (see Section 4.7 and 5.2). As in the circular case, after the spherical phase error has exceeded half a wavelength, the copolar patterns tend towards a constant beamwidth and for large flare angles the beamwidth is determined principally by the flare angle and not the aperture size.

Reference has already been made to the need for particular care in designing the throat section of a rectangular corrugated horn. The principle of design is similar to the circular case, namely, the first slot should be half a guidewavelength deep. The fact that the HE_{12} mode has a TE_{12}/TM_{12} type field at cut-off and is not the dominant mode in the rectangular waveguide poses additional problems. The half guide wavelength slot should be chosen at the highest operating frequency so that it is never deeper than $\lambda_g/2$. The slot depths can then be tapered from $\lambda_g/2$ to $\lambda_g/4$ over a few wavelengths. The throat size should be as large as feasible because the modes propagating in the deep slots are very dispersive (in contrast to the circular horn). This means that the VSWR characteristics are also highly frequency dependent, although the large throat size means that the absolute level can be kept low. The maximum size of the throat is determined by the need to avoid higher order mode excitation. Unfortunately, the waveguide fields of the next HE/EH modes above the HE_{12} mode have similar patterns to the HE_{12} mode. This means that higher order modes can be easily excited. Add to this difficulty the approximate nature of the theory and the difficulties of manufacturing rectangular corrugated horns and it is easy to see why they have not attracted much interest in the past. The only reason to choose a rectangular structure rather than a circular structure is the ability to generate elliptical radiation patterns. For all other criteria the circular horn is superior.

7.3 Elliptical corrugated horns

Elliptical corrugated waveguides have recieved relatively little attention to date. This is partly due to the difficulties of analysis and partly due to the problems of manufacturing an elliptical corrugated horn. However, the need to produce elliptically contoured beams with high polarisation purity is increasing the interest in these horns. Because this type of horn is still being studied we will confine ourselves to stating the fundamentals of the analysis of elliptical corrugated waveguides.

The first work on elliptical corrugated horns appears to have been done by Jeuken *et al.* [78, 83] who used a simple analytical model and also built a horn with good circular polarisation characteristics. The horn was made by machining. The crosspolar performance of a horn made by a series of elliptical discs was reported by Guy and Ashton [67]. They found that the peak crosspolarisation when the electric field was parallel to the minor axis of the ellipse was below -35 dB over a 35% band but when the electric field was parallel to the major axis of the ellipse this band was reduced to only a few percent. A detailed analysis of the elliptical corrugated waveguide has been given by Al-Hariri [4] who studied the low attenuation properties of the waveguide.

A sketch of the elliptical corrugated waveguide is shown in Fig. 7.7(a). The elliptic cylinder co-ordinate system (ξ, η, z) is shown in Fig. 7.7(b), where the contour surfaces of constant ξ are confocal elliptic cylinders, and those of constant η are confocal hyperbolic cylinders. The constant h represents half the distance between the foci. The confocal cylinder $\xi = \xi_0$ coincides with the boundary of the waveguide, and the z-axis coincides with the longitudinal axis of the elliptic waveguide. The eccentricity e of the cross-section is given by $1/\cosh \xi_0$. The major and minor axes are $2a = 2h \cosh \xi_0$ and $2b = 2h \sinh \xi_0$, respectively.

The wave equation in the elliptic cylinder system has the form:

$$\frac{\partial^2 \Lambda}{\partial \xi^2} + \frac{\partial^2 \Lambda}{\partial \eta^2} 2q(\cosh 2\xi - \cos 2\eta)\,\Lambda = 0 \tag{7.29}$$

where $4q = K^2 h^2$ and $K^2 = \omega^2 \epsilon_0 \mu_0 - \beta^2$.

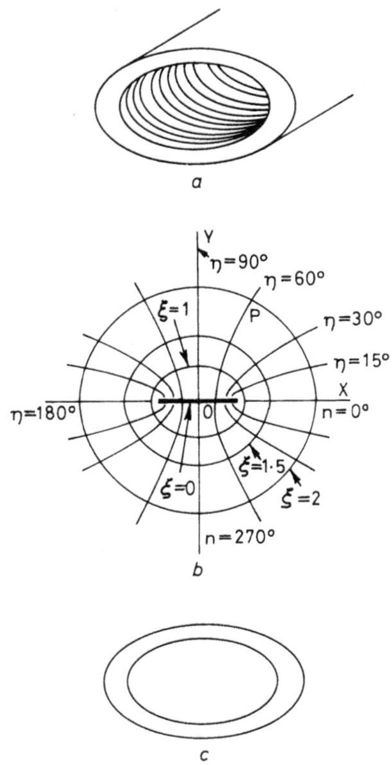

Fig. 7.7 *Elliptical corrugated waveguide*
a Waveguide
b Elliptical co-ordinate system consisting of confocal ellipses and hyperbolas. P has co-ordinates $\xi = 2$, $\eta = \pi/3$
c Two ellipses with the same eccentricity, showing varying slot depth with η

In order to obtain solutions of eqn. 7.29 we set:

$$\Lambda(\xi, \mu) = R(\xi)\,\theta(\eta) \tag{7.30}$$

Substituting into eqn. 7.29 and applying the usual separation of variable procedure, two ordinary differential equations are obtained:

$$\frac{d^2\theta(\eta)}{d\eta^2} + (C - 2q\cos 2\eta)\theta(\eta) = 0 \tag{7.31}$$

and

$$\frac{d^2R(\xi)}{d\xi^2} - (C - 2q \cosh 2\xi)R(\xi) = 0 \tag{7.32}$$

where C is the separation constant.

Eqn. 7.31 is the angular Mathieu differential equation; eqn. 7.32 which follows from eqn. 7.31 by the transformation $\eta = \pm j\xi$, is the modified Mathieu differential equation.

For a physically admissible single-valued electromagnetic field, $\Lambda(\xi, \eta)$ must be a periodic function of η, of period π or 2π. The separation constant C, which in this case must be a function of q, is an infinite set of characteristic values for every q. When q is real the characteristic values are real; when q is negative imaginary, as it is in one of the cases which is considered, the characteristic values can be either real or complex.

Corresponding to $q = 0$ there are two independent periodic solutions, namely $\sin(v\eta)$ and $\cos(v\eta)$ with the separation constant $C = v^2$, where v is an integer. It can be shown that when q differs from zero, a characteristic value C determines one and only one periodic solution which is either even or odd in η. The characteristic value C, giving rise to even and odd solutions are denoted by $a_v(q)$ and $b_v(q)$, respectively. The subscript v identifies those sets of characteristic values which approach v^2 as $q \to 0$.

For an arbitrary positive real q, the periodic solutions of Mathieu eqn. 7.31 are:

$$\theta(\eta) \begin{cases} = ce_v(\eta, q) & \text{even } a_v(q) \\ = se_v(\eta, q) & \text{odd } b_v(q) \end{cases} \tag{7.33}$$

where $ce_v(\eta, q)$ and $se_v(\eta, q)$ are, respectively, the even and odd angular Mathieu functions. The corresponding solutions for the modified Mathieu eqn. 7.32 are

$$R(\xi) \begin{cases} = L_1 Ce_v(\xi, q) + L_2 Fey_v(\xi, q) & \text{even } a_v(a) \\ = L_3 Se_v(\xi, q) + L_4 Gey_v(\xi, q) & \text{odd } b_v(q) \end{cases} \tag{7.34}$$

where $Ce_v(\xi, q)$ and $Se_v(\xi, q)$ are radial even and odd Mathieu functions of the first kind, $Fey_v(\xi, q)$ and $Gey_v(\xi, q)$ are radial even and odd Mathieu functions of the second kind. L_1, L_2, L_3 and L_4 are constants.

The fields in the inner region of the corrugated waveguide for modes whose axial magnetic field component is represented by the even Mathieu functions are then given by:

$$
\begin{aligned}
{}_1E_z &= \sum_{v=0} B_v Se_v(q_1, \xi) se_v(q_1, \eta) \\
{}_1H_z &= Y_0 \sum_{v=1} A_v Ce_v(q_1, \xi) ce_v(q_1, \eta) \\
{}_1E_\eta &= -\mathrm{j}\frac{\beta}{lK^2} \sum_v B_v Se_v(q_1, \xi) se'_v(q_1, \eta) \\
&\quad + \mathrm{j}\frac{k}{lK^2} \sum_v A_v Ce'_v(q_1, \xi) ce_v(q_1, \eta) \\
{}_1H_\eta &= -\mathrm{j}\frac{\omega\epsilon_0}{lK^2} \sum_v B_v Se'_v(q_1, \xi) se_v(q_1, \eta) \\
&\quad - \mathrm{j}\frac{\beta Y_0}{lK^2} \sum_v A_v Ce_v(q_1, \xi) ce'_v(q_1, \eta) \\
{}_1E_\xi &= -\mathrm{j}\frac{\beta}{lK^2} \sum_v B_v Se'_v(q_1, \xi) se_v(q_1, \eta) \\
&\quad - \mathrm{j}\frac{k}{K^2} \sum_v A_v Ce_v(q_1, \xi) ce'_v(q_1, \eta) \\
{}_1H_\xi &= \mathrm{j}\frac{\omega\epsilon_0}{lK^2} \sum_v B_v Se_v(q_1, \xi) se'_v(q_1, \eta) \\
&\quad - \mathrm{j}\frac{\beta Y_0}{lK^2} \sum_v A_v Ce'_v(q_1, \xi) ce_v(q_1, \eta)
\end{aligned}
\tag{7.35}
$$

The factor $\exp \mathrm{j}(\omega t - \beta z)$ is assumed throughout, and

$$l = h(\cosh^2 \xi - \cos^2 \eta)^{1/2}$$

These equations are derived under the assumption that there are a large number of corrugations per wavelength and that the higher order space harmonics can be ignored.

The fields in the corrugations will be assumed to be those of a cut-off TM type mode. One of the complications of the analysis of elliptical corrugated waveguides is that the ellipse which forms the outer boundary of the waveguide has a different eccentricity from the ellipse which forms the inner boundary. This can be seen from Fig. 7.7(c). If the eccentricities of the two ellipses were identical then the slot depth would vary around the perimeter. Thus a separate set of parameters q_2, ξ_2 and eigennumber r are needed in the outer region.

The field components in the outer region are then given by:

$$
\begin{aligned}
{}_2E_z &= \sum_{r=1} L_r \Pi_r(q_2, \xi_2, \xi) se_r(q_2, \eta) \\
{}_2H_\eta &= \mathrm{j}\,\frac{\omega\epsilon_0}{lk^2} \sum_r L_r \Pi_r'(q_2, \xi_2, \xi) se_r'(q_2, \eta) \\
{}_2H_\xi &= \mathrm{j}\,\frac{\omega\epsilon_0}{lk^2} \sum_r L_r \Pi_r(q_2, \xi_2, \xi) se_r'(q_2, \eta)
\end{aligned}
\tag{7.36}
$$

where $4q_1 = h^2(k^2 - \beta^2)$, $4q_2 = h^2k^2$, $k^2 = \omega^2\epsilon_0\mu_0$.

In the above equation

$$
\Pi_i(q, \xi_2, \xi) = \frac{Se_i(q, \xi) Gey_i(q, \xi_2) - Se_i(q, \xi_2) Gey_i(q, \xi)}{Gey_i(q, \xi_2)} \tag{7.37}
$$

The characteristic equation is obtained in the usual manner by matching the fields at the boundary $\xi = \xi_1$. The equation must be derived assuming an infinite set of eigennumbers v and r. This leads to considerable complexity in the analysis and is due to the presence of the angular Mathieu functions which are not only functions of η, the angular co-ordinate, but also of the electrical properties of the medium in which they apply. The impedance of the slot region ${}_2E_z/{}_2H_\eta$ is a function of the aximuthal position (η) around the waveguide. This is in contrast to the circular waveguide where the impedance is independent of the azimuthal co-ordinate.

After manipulation the characteristic equation for even HE modes with only the first term in the v summation becomes:

$$
\left[Se_1'(q_1, \xi_1) - \left(\frac{K}{k}\right)^2 \frac{\Pi_1(q_2, \xi_1, \xi_2)}{\Pi_1(q_2, \xi_1, \xi_2)}\, Se_1(q_1, \xi_1)\right] \theta_{1,1}
+ \bar{\beta}^2 \sum_r Se_1(q_1, \xi_1) \frac{Ce_1(q_1, \xi_1)}{Ce_r'(q_1, \xi_1)} \alpha_{1,r}\,\delta_{r,1} = 0 \tag{7.38}
$$

where

$$
\begin{aligned}
\theta_{v,\mu} &= \int_0^{2\pi} se_v(q_1, \eta) se_\mu(q_2, \eta)\,\mathrm{d}\eta \Big/ \int_0^{2\pi} se_\mu^2(q_2, \eta)\,\mathrm{d}\eta \\
\alpha_{v,\mu} &= \int_0^{2\pi} se_{v'}(q_1, \eta) ce_\mu(q_1, \eta)\,\mathrm{d}\eta \Big/ \int_0^{2\pi} ce_\mu^2(q_1, \eta)\,\mathrm{d}\eta \\
\delta_{v,\mu} &= \int_0^{2\pi} ce_{v'}(q_1, \eta) se_\mu(q_2, \eta)\,\mathrm{d}\eta \Big/ \int_0^{2\pi} se_\mu^2(q_2, \eta)\,\mathrm{d}\eta
\end{aligned}
$$

A similar equation can be obtained for odd type Mathieu functions which then leads to odd type HE_{11} modes. If the elliptical waveguide is made circular (eccentricity = 0), then eqn. 7.38 becomes the familiar characteristic equation for circular corrugated waveguides.

The computation of the characteristic equation is time consuming because the Mathieu functions are difficult to calculate. The angular Mathieu functions may be expanded in terms of an infinite series of trigonometric functions and the modified Mathieu functions may be expanded in terms of an infinite series of products of Bessel functions [178]. Computation of the Mathieu functions involves finding the root of an infinite continued-fraction transcendental equation and then using a recurrence relation.

Typical dispersion characteristics are shown in Fig. 7.8 for even and odd HE_{11} modes. The two modes are degenerate at the frequency where the slot depth is approximately a quarter of a wavelength deep.

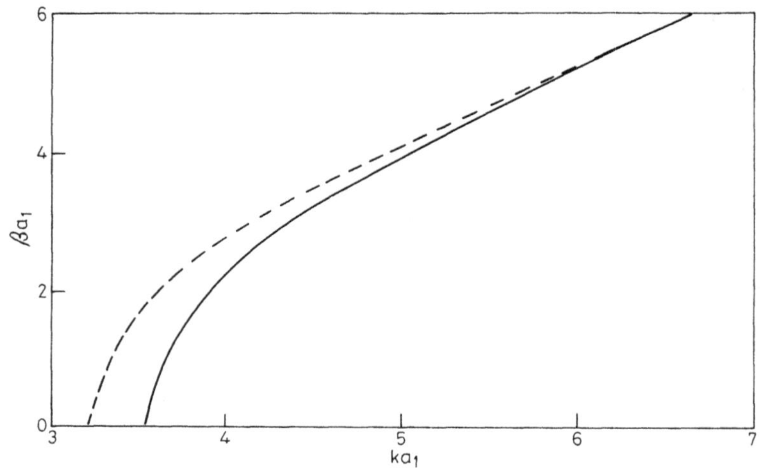

Fig. 7.8 *Dispersion curves for HE_{11} mode, $e_1 = 0{\cdot}7$, $e_2 = 0{\cdot}56$*
——— odd mode
– – – – even mode

The radiation characteristics of elliptical corrugated waveguides can be derived by Fourier transforming the aperture fields to give the far-fields. Jansen and Jeuken [78] have done this under the assumption that the impedance of the slots are $Z_\eta = 0$ and $Z = \infty$, and are independent of frequency. They found that the far-field is circularly polarised everywhere, if the even and odd HE_{11} modes in the aperture had a phase difference of 90°. They also found that the even mode and the corresponding odd mode give rise to the same radiation pattern.

Appendix

Cylindrical corrugated waveguide analysis including space harmonics

Introduction

In this appendix the exact mathematical analysis including space harmonics is presented for cylindrical corrugated waveguides.

The propagation equation, giving a solution for angular frequency ω in terms of the phase change coefficient β, is derived by the usual technique for waveguides with two or more regions. That is, the waveguide is divided into the inner region $\mathrm{I}(r < r_1)$ and the outer region $\mathrm{II}(r < r_1)$ (slot region). The field expansions which satisfy Maxwell's equations in the two regions are matched at the common boundary $r = r_1$ and the field coefficients eliminated so as to leave a propagation equation determinant. This transcendental determinant is equated to zero and the correct $\beta : \omega$ relationship computed by an iterative process.

The field coefficients and hence electric and magnetic field components are obtained by back substitution using the propagation determinant.

The geometry to be analysed is shown in Fig. 3.2. The waveguide is assumed to be infinite in length. In this appendix we adopt a notation widely used in waveguide analysis and call the azimuthal order n, whereas elsewhere in this book m has been used to follow convention in papers dealing specifically with horns.

The slots will support a TM_{n0} mode (E_z, E_r, H_ψ) and an infinite set of TM_{nm} and TE_{nm} standing waves which satisfy the boundary conditions at $z = \pm(b/2)$ and $r = r_0$. In practice the infinite set is truncated at a finite value L so that

$$\begin{matrix} E_z \\ E_r \\ H_\Psi \end{matrix} = \sum_{m=0}^{L} \begin{matrix} E_{z_m} \\ E_{r_m} \\ H_{\Psi_m} \end{matrix}$$

$$\begin{matrix} H_z \\ H_r \\ E_\Psi \end{matrix} = \sum_{m=1}^{L} \begin{matrix} H_{z_m} \\ H_{r_m} \\ E_{\Psi_m} \end{matrix} \qquad (1)$$

The simple surface impedance theory assumes that only the TM_{n0} mode is present in the slot, i.e. $L = 0$ in this analysis. This mode has no z dependence in the slot.

In periodic structures, Floquet's therem states that it is sufficient to consider the fields in just one slot or unit cell. The fields in any other unit cell, distance Qp in the z direction (Q interger) from the basic cell are $e^{j\beta_0 Qp}$ times those in the basic cell where $\beta_0 p$ is the phase change per unit cell.

In the inner region, $r < r_1$, the boundary conditions to be satisfied are that the tangential electric field and the normal magnetic flux density vanish at $r = r_1$ for $b/2 \leqslant |z| \leqslant p/2$. All fields over the plane $z = p/2$ will be $e^{j\beta_0 p}$ times the corresponding fields at $z = -(p/2)$. Each fundamental mode in the inner region has associated with it a double infinite set of space harmonics. As in the case of the outer region the series is restricted to a finite value, $\pm K$, so that

$$E_z = \sum_{N=-K}^{+K} E_{zN} \tag{2}$$

and similarly for all other field components.

The E_z and H_z fields are derived from the wave equation and the other field components from:

$$\begin{aligned}
E_\Psi &= \frac{1}{K^2 r}\frac{\partial^2 Ez}{\partial\Psi\partial z} + \frac{j\omega\mu_0}{K^2}\frac{\partial Hz}{\partial r} \\
H_\Psi &= \frac{1}{K^2 r}\frac{\partial^2 Hz}{\partial z^2} - \frac{j\omega\epsilon_0}{K^2}\frac{\partial Ez}{\partial r} \\
E_r &= \frac{1}{K^2}\frac{\partial^2 Ez}{\partial\Psi\partial z} - \frac{j\omega\mu_0}{K^2 r}\frac{\partial Hz}{\partial\Psi} \\
H_r &= \frac{1}{K^2}\frac{\partial^2 Hz}{\partial r\partial z} + \frac{j\omega\epsilon_0}{r}\frac{\partial Ez}{\partial\Psi}
\end{aligned} \tag{3}$$

All fields are assumed to have a time dependence $e^{j\omega t}$.

Field components in corrugated waveguide

Region I (inner) $r < r_1$

$$E_{zN} = A_N J_n(K_N r) e^{-j\beta_N z} \cos n\Psi \tag{4a}$$

$$H_{zN} = y_0 B_N J_n(K_N r)\, e^{-j\beta_N z} \sin n\Psi \tag{4b}$$

$$E_{\Psi N} = j\left[A_N\, n\, \frac{\beta_N r}{(K_N r)^2} J_n(K_N r) + B_N \frac{\bar{r}}{(K_N r)} J_n'(K_N r)\right] e^{-j\beta_N z} \sin n\Psi \tag{4c}$$

$$H_{\Psi_N} = -\mathrm{j}y_0\left[A_N \frac{\bar{r}}{(K_N r)} J_n'(K_N r) + B_N \frac{\beta_N r}{(K_N r)^2} n J_n(K_N r)\right] \mathrm{e}^{-\mathrm{j}\beta_N z}\cos n\Psi \tag{4d}$$

$$E_{r_N} = -\mathrm{j}\left[A_N \frac{\beta_N r}{(K_N r)} J_n'(K_N r) + B_N \frac{\bar{r}}{(K_N r)^2} n J_n(K_N r)\right] \mathrm{e}^{-\mathrm{j}\beta_N z}\cos n\Psi \tag{4e}$$

$$H_{r_N} = -\mathrm{j}y_0\left[A_N \frac{\bar{r}}{(K_N r)^2} n J_n(K_N r) + B_N \frac{(\beta_N r)}{(K_N r)} J_n'(K_N r)\right] \mathrm{e}^{-\mathrm{j}\beta_N z}\sin n\Psi \tag{4f}$$

Region II (outer) $r > r_1$

$$E_{z_m} = [c_m S_n^D \cos\eta_m z - \mathrm{j}c_m' S_n^{D'} \sin\eta_m' z]\cos n\Psi \tag{5a}$$

$$H_{z_m} = y_0[-\mathrm{j}d_m R_n^D \sin\eta_m z + d_m' R_n^{D'} \cos\eta_m' z]\sin n\Psi \tag{5b}$$

$$E_{\Psi_m} = \left\{\left[c_m n \frac{\eta_m r}{(\Gamma_m r)^2} S_n^D + d_m \frac{\bar{r}}{(\Gamma_m r)} R_n^N\right]\sin\eta_m z + \mathrm{j}\left[c_m' n \frac{\eta_m' r}{(\Gamma_m' r)^2} S_n^{D'} + d_m' \frac{\bar{r}}{(\Gamma_m' r)} R_n'\right]\cos\eta_m' z\right\}\sin n\Psi \tag{5c}$$

$$H_{\Psi_m} = y_0\left\{-\mathrm{j}\left[d_m n \frac{\eta_m r}{(\Gamma_m r)^2} R_n^D + c_m \bar{r}\frac{(\Gamma_m r)}{(\Gamma_m r)^2} S_n^{N'}\right]\cos\eta_m z - \left[d_m' n \frac{\eta_m' r}{(\Gamma_m' r)^2} R_n^{D'} + c_m' \bar{r}\frac{(\Gamma_m' r)}{(\Gamma_m' r)^2} S_n^{N'}\right]\sin\eta_m' z\right\}\cos n\Psi \tag{5d}$$

$$E_{r_m} = -\left\{\left[c_m \frac{\eta_m r}{(\Gamma_m r)^2} S_n^N + d_m n \frac{\bar{r}}{(\Gamma_m r)^2} R_n^D\right]\sin\eta_m z + \mathrm{j}\left[c_m' \frac{\eta_m' r}{(\Gamma_m' r)^2} S_n^{N'} + d_m' n \frac{\bar{r}}{(\Gamma_m' r)^2} R_n^{D'}\right]\cos\eta_m' z\right\}\cos n\Psi \tag{5e}$$

$$H_{r_m} = -y_0\left\{\mathrm{j}\left[c_m \frac{n\bar{r}}{(\Gamma_m r)^2} S_n^D + d_m \frac{\eta_m r}{(\Gamma_m r)} R_n^N\right]\cos\eta_m z + \left[c_m' \frac{n\bar{r}}{(\Gamma_m' r)^2} S_n^{D'} + d_m' \frac{\eta_m' r}{(\Gamma_m' r)} R_n^{N'}\right]\sin\eta_m' z\right\}\sin n\Psi$$

In the above equations the symbols have the following meanings:

$$\beta_N = \beta_0 + \frac{2\pi N}{p}$$

$$(K_N r)^2 = \bar{r}^2 - \beta_N^2 r^2$$

$$\bar{r} = \frac{\omega r}{c} = \frac{2\pi r}{\lambda_0}$$

$$\eta_m r = \frac{2m\pi}{b} r$$

$$\eta'_m r = (2m-1)\frac{\pi}{b} r$$

$$(\Gamma_m r)^2 = \bar{r}^2 - (\eta_m r)^2$$

$$(\Gamma'_m r)^2 = \bar{r}^2 - (\eta'_m r)^2$$

$$R_n^N = J'_n(\Gamma_m r)\, Y'_n(\Gamma_m r_0) - J'_n(\Gamma_m r_0)\, Y'_n(\Gamma_m r)$$

$$R_n^D = J_n(\Gamma_m r)\, Y'_n(\Gamma_m r_0) - J'_n(\Gamma_m r_0)\, Y_n(\Gamma_m r)$$

$$S_n^N = J'_n(\Gamma_m r)\, Y_n(\Gamma_m r_0) - J_n(\Gamma_m r_0)\, Y'_n(\Gamma_m r)$$

$$S_n^D = J_n(\Gamma_m r)\, Y_n(\Gamma_m r_0) - J_n(\Gamma_m r_0)\, Y_n(\Gamma_m r)$$

$R_n^{N'}, R_n^{D'}, S_n^{N'}, S_n^{D'}$, $(\Gamma'_m r)$ replaces $(\Gamma_m r)$.

In the above and in eqns. 4 and 5, when $K_N r$, $\Gamma_m r$ or $\Gamma'_m r$ are imaginary, $J_n(x)$ is replaced by $I_n(x)$ and $Y_n(x)$ by $K_n(x)$.

A_N, B_N coefficients of E and H mode components in inner region

c_m, c'_m coefficients of E mode components in outer region

d_m, d'_m coefficients of H mode components in outer region

y_0 admittance of free space

Propagation equation

The fields in regions I and II are matched at $r = r_1$ by equating the z and Ψ components of the total fields at the interface $r = r_1$. The E_z and E_Ψ field components in the slot $-(b/2) < z < (b/2)$ are equated to the E_z and E_Ψ fields over the length $-(b/2) < z < (b/2)$ in the inner region and the orthogonality of the $e^{-j\beta_N z}$ functions used to give the coefficients of any space harmonic in the inner region in terms of all the coefficients in the slot region.

Similarly, the H_z and H_Ψ field components are equated over the length $-(b/2) < z < (b/2)$ and the orthogonality of the sin $(\eta_m z)$ and cos $(\eta_m z)$ functions used to give the coefficients of any mode in the slot region in terms of all the coefficients in the inner region.

Hence:

$$\int_{-p/2}^{p/2} E_z^{\mathrm{I}} e^{j\beta_N z} dz = \int_{-b/2}^{b/2} E_z^{\mathrm{II}} e^{j\beta_N z} \tag{6}$$

$$\int_{-p/2}^{p/2} E_\Psi^{\mathrm{I}} e^{j\beta_N z} dz = \int_{-b/2}^{b/2} E_\Psi^{\mathrm{II}} e^{j\beta_N z} dz \tag{7}$$

$$\int_{-b/2}^{b/2} H_z^{\mathrm{II}} \begin{matrix}\cos \eta_m' z\\ \sin \eta_m z\end{matrix} dz = \int_{-b/2}^{b/2} H_z^{\mathrm{I}} \begin{matrix}\cos \eta_m' z\\ \sin \eta_m z\end{matrix} dz \tag{8}$$

$$\int_{-b/2}^{b/2} H_\Psi^{\mathrm{II}} \begin{matrix}\cos \eta_m z\\ \sin \eta_m' z\end{matrix} dz = \int_{-b/2}^{b/2} H_\Psi^{\mathrm{I}} \begin{matrix}\cos \eta_m z\\ \sin \eta_m' z\end{matrix} dz \tag{9}$$

The orthogonality integrals are on the left-hand side while the integrals on the right-hand side are of the form:

$$I_1 = \int_{-b/2}^{b/2} \cos(\eta_m z) e^{j\beta_N z} dz$$

$$= \tfrac{1}{2} \int_{-b/2}^{b/2} [\cos(\eta_m - \beta_N)z + \cos(\eta_m + \beta_N)z]\, dz$$

$$= \frac{\sin\left(\dfrac{\beta_N b}{2} - \dfrac{\eta_m b}{2}\right)}{(\beta_N - \eta_m)} + \frac{\sin\left(\dfrac{\beta_N b}{2} + \dfrac{\eta_m b}{2}\right)}{(\beta_N + \eta_m)}$$

or

$$I_1 = b\left(\frac{\beta_N b}{2}\right)(-1)^m \frac{\sin(\beta_N b/2)}{(\beta_N b/2)^2 - (\pi m)^2} \tag{10}$$

Similarly:

$$+jI_2 = \int_{-b/2}^{b/2} \sin'_m z\, e^{-j\beta_N z} dz = jb\left(\frac{\beta_N b}{2}\right)(-1)^m$$

$$\frac{\cos(\beta_N b/2)}{(\beta_N b/2)^2 - (\pi m - \pi/2)^2} \tag{11}$$

$$-jI_3 = \int_{-b/2}^{b/2} \sin \eta_m z\, e^{-j\beta_N z} dz = jb(\pi m)(-1)^m$$

$$\frac{\sin(\beta_N b/2)}{(\beta_N b/2)^2 - (\pi m)^2} \tag{12}$$

$$I_4 = \int_{-b/2}^{b/2} \cos \eta'_m z \, e^{-j\beta_N z} = b(\pi m - \pi/2) \frac{(-1)^m \cos(\beta_N b/2)}{(\beta_N b/2)^2 - (\pi m - \pi/2)^2} \tag{13}$$

Then substituting the field components of eqns. 4 and 5 into eqns. 6 to 9 gives:

E_z components:

$$A_N p J_n(K_N r_1) = \sum_{m=0}^{L} [c_m S_n^D I_1 + c'_m S_n^{D'} I_2] \tag{14}$$

for $N = 0, \pm 1, \pm 2, \ldots \pm K$.

E_Ψ components:

$$\left[A_N n \frac{(\beta_N r_1)}{(K_N r_1)^2} J_n(K_N r_1) + B_N \frac{\bar{r}_1}{(K_N r_1)} J'_n(K_N r_1)\right] p$$

$$= \sum_{m=1}^{L} \left\{ \left[c_m n \frac{(\eta_m r_1)}{(\Gamma_m r_1)^2} S_n^D + d_m \frac{\bar{r}_1}{(\Gamma_m r_1)} R_n^N\right] I_3 \right. \tag{15}$$

$$\left. + \left[c'_m n \frac{(\eta'_m r_1)}{(\Gamma'_m r_1)} S_n^{D'} + d'_m \frac{\bar{r}_1}{(\Gamma'_m r_1)} R_n^{N'}\right] I_4 \right\}$$

for $N = 0, \pm 1, \pm 2, \ldots \pm K$.

Real and imaginary H_Ψ components:

$$\frac{b}{2}\left[c_m \frac{\bar{r}_1}{\Gamma_m r_1} S_n^N + d_m n \frac{(\eta_m r_1)}{(\Gamma_m r_1)^2} R_n^D\right]$$

$$= \sum_{I=-K}^{K} \left[B_I \frac{(\beta_I r_1)}{(K_I r_1)^2} n J_n(K_I r_1) + A_I \frac{\bar{r}_1}{K_I r_1} J'_n(K_I r_1)\right] I_1 \tag{16}$$

for $m = 0, 1, 2 \ldots L$.

$$\frac{b}{2}\left[c'_m \frac{\bar{r}_1}{\Gamma'_m r_1} S_n^{N'} + d'_m n \frac{(\eta'_m r_1)}{(\Gamma'_m r_1)^2} R_n^{D'}\right]$$

$$= \sum_{I=-K}^{K} \left[B_I \frac{(\beta_I r_1)}{(K_I r_1)^2} n J_n(K_I r_1) + A_I \frac{\bar{r}_1}{K_I r_1} J'_n(K_I r_1)\right] I_2 \tag{17}$$

for $m = 1, 2, \ldots L$.

Real and imaginary H_z components:

$$\frac{b}{2} d_m R_n^D = \sum_{I=-K}^{K} B_I J_n(K_I r_1) I_3 \tag{18}$$

$$\frac{b}{2} d'_m R_n^{D'} = \sum_{I=-K}^{K} B_I J_n(K_I r_1) I_4 \tag{19}$$

both for $m = 1, 2, \ldots L$.

Eqn. 16, left-hand side, should be multiplied by two for $m = 0$. Substituting eqns. 16 to 19 into eqns. 14 and 15 gives two sets of simultaneous equations for the As and Bs:

$$\begin{aligned} \sum_{I=-K}^{+K} [U_{NI} A_I + V_{NI} B_I] &= 0 \\ \sum_{I=-K}^{+K} [X_{NI} A_I + Y_{NI} B_I] &= 0 \end{aligned} \tag{20}$$

for $N = 0, \pm 1, \pm 2, \ldots . \pm K$, where

$$U_{IN} = \delta_{IN} \frac{p}{b} J_n(K_N r_1) - \sum_{m=0}^{L} L_{NI_m} \tag{21}$$

$$V_{NI} = -\sum_{m=0}^{L} M_{NI_m} \tag{22}$$

$$X_{NI} = \delta_{IN} \frac{p}{b} n \frac{(\beta_N r_1)}{(K_N r_1)^2} J_n(K_N r_1) - \sum_{m=1}^{L} P_{NI_m} \tag{23}$$

$$Y_{NI} = \delta_{IN} \frac{p}{b} \frac{\bar{r}_1}{(K_N r_1)} J'_n(K_N r_1) - \sum_{m=1}^{L} Q_{NI_m} \tag{24}$$

δ_{IN} is the kronecker delta, $= 1$ for $I = N$; $= 0$ for $I \neq N$.

$$\begin{aligned} \sum L_{NI_m} = \frac{J'_n(K_I r_1)}{(K_I r_1)} (\beta_I b/2) \Bigg[\Bigg\{ & \sin(\beta_I b/2)(\beta_N b/2) \sin(\beta_N b/2) \\ & \sum_{m=0}^{L} \left[\epsilon_m \frac{S_n^D}{S_n^D} (\Gamma_m r_1) D_{I_m} D_{N_m} \right] \Bigg\} \\ & + \Bigg\{ 2 \cos(\beta_I b/2)(\beta_N b/2) \cos(\beta_N b/2) \\ & \sum_{m=1}^{L} \left[\frac{S_n^{D'}}{S_n^{N'}} (\Gamma'_m r_1) D'_{I_m} D'_{N_m} \right] \Bigg\} \Bigg] \end{aligned} \tag{25}$$

$$\sum M_{NI_m} = \frac{J_n(K_I r_1)}{\bar{r}_1} \, n \frac{(\beta_I b/2)(\beta_I r_1)}{(K_I r_1)^2} \sin\left(\frac{\beta_I b}{2}\right)\left(\frac{\beta_N b}{2}\right)$$

$$\sin\left(\frac{\beta_N b}{2}\right)\sum_{m=0}^{L}\left[\epsilon_m \frac{S_n^D}{S_n^N}(\Gamma_m r_1) D_{I_m} D_{N_m}\right]\Bigg\}$$

$$+\left\{2\cos\left(\frac{\beta_I b}{2}\right)\left(\frac{\beta_N b}{2}\right)\cos\left(\frac{\beta_N b}{2}\right)\sum_{m=1}^{L}\left[\frac{S_n^{D'}}{S_n^{N'}}(\Gamma'_m r_1) D'_{I_m} D'_{N_m}\right]\right\}$$

$$-\left\{2\sin\left(\frac{\beta_I b}{2}\right)\left(\frac{\beta_N b}{2}\right)\cos\left(\frac{\beta_N b}{2}\right)\sum_{m=1}^{L}\left[\frac{S_n^D}{S_n^N}\,\frac{(\pi m)(\eta_m r_1)}{(\Gamma_m r_1)} D_{I_m} D_{N_m}\right]\right\}$$

$$-\left\{2\cos\left(\frac{\beta_I b}{2}\right)\left(\frac{\beta_N b}{2}\right)\cos\left(\frac{\beta_N b}{2}\right)\right.$$

$$\left.\sum_{m=1}^{L}\frac{S_n^{D'}}{S_n^{N'}}\,\frac{\pi(m-\frac{1}{2})(\eta'_m r_1)}{(\Gamma'_m r_1)} D'_{I_m} D'_{N_m}\right\}\Bigg] \qquad (26)$$

$$P_{NI_m} = 2n\frac{J'_n(K_I r_1)}{(K_I r_1)}\left(\frac{\beta_I b}{2}\right)\left[\left\{\sin\left(\frac{\beta_I b}{2}\right)\sin\left(\frac{\beta_N b}{2}\right)\right.\right.$$

$$\left.\sum_{m=1}^{L}\left[\frac{S_n^D}{S_n^N}\,\frac{(\pi m)(\eta_m r_1)}{(\Gamma_m r_1)} D_{I_m} D_{N_m}\right]\right\} + \left\{\cos\left(\frac{\beta_I b}{2}\right)\cos\left(\frac{\beta_N b}{2}\right)\right.$$

$$\left.\left.\sum_{m=1}^{L}\left[\frac{S^{D'}}{S_n^{N'}}\,\frac{\pi(m-\frac{1}{2})(\eta'_m r_1)}{(\Gamma'_m r_1)} D'_{I_m} D'_{N_m}\right]\right\}\right] \qquad (27)$$

$$\sum Q_{NI_m} = 2\frac{J_n(K_I r_1)}{\bar{r}_1 (K_I r_1)^2}\, n^2 \left(\frac{\beta_I b}{2}\right)(\beta_I r_1)$$

$$\left\{\sin\left(\frac{\beta_I b}{2}\right)\sin\left(\frac{\beta_N b}{2}\right)\sum_{m=1}^{L}\left[\frac{S_n^D}{S_n^N}\,\frac{(\pi m)(\eta_m r_1)}{(\Gamma_m r_1)} D_{I_m} D_{N_m}\right]\right.$$

$$+\cos\left(\frac{\beta_I b}{2}\right)\cos\left(\frac{\beta_N b}{2}\right)$$

$$\left.\sum_{m=1}^{L}\left[\frac{S_n^{D'}}{S_n^{N'}}\,\frac{\pi(m-\frac{1}{2})(\eta'_m r_1)}{(\Gamma'_m r_1)} D_{I_m} D_{N_m}\right]\right\}$$

$$-2n^2\frac{J_n(K_I r_1)}{\bar{r}_1}\left\{\sin\left(\frac{\beta_I b}{2}\right)\sin\left(\frac{\beta_N b}{2}\right)\right.$$

$$\sum_{m=1}^{L}\left[\frac{S_n^D}{S_n^N}\,\frac{(\pi m)^2(\eta_m r_1)^2}{(\Gamma_m r_1)^3}\,D_{I_m}D_{N_m}\right]$$

$$+\cos\left(\frac{\beta_I b}{2}\right)\sin\left(\frac{\beta_N b}{2}\right)\sum_{m=1}^{L}\left[\frac{S_n^{D'}}{S_n^{N'}}\,\frac{\pi^2(m-\frac{1}{2})^2(\eta_m' r_1)^2}{(\Gamma_m' r_1)^3}\,D'_{I_m}D'_{N_m}\right]\Bigg\}$$

$$+\,2\,J_n(K_I r_1)\bar{r}_1\left\{\sin\left(\frac{\beta_I b}{2}\right)\sin\left(\frac{\beta_N b}{2}\right)\right.$$

$$\sum_{m=1}^{L}\left[\frac{R_n^N}{R_n^D}\,\frac{(\pi m)^2}{(\Gamma_m r_1)}\,D_{I_m}D_{N_m}\right]$$

$$+\cos\left(\frac{\beta_I b}{2}\right)\cos\left(\frac{\beta_N b}{2}\right)\sum_{m=1}^{L}\left[\frac{R_n^{N'}}{R_n^{D'}}\,\frac{\pi^2(m-\frac{1}{2})^2}{(\Gamma_m' r_1)}\,D'_{I_m}D'_{N_m}\right]\Bigg\} \quad (28)$$

$\epsilon_m = 1$ for $m = 0$; $= 2$ for $m \neq 0$.

In the above equations, the integrals $I_1 - I_4$ have been written for convenience as:

$$I_{1_N} = b\left(\frac{\beta_N b}{2}\right)\sin\left(\frac{\beta_N b}{2}\right)D_{N_m}$$

$$I_{2_N} = b\left(\frac{\beta_N b}{2}\right)\cos\left(\frac{\beta_N b}{2}\right)D'_{N_m}$$

$$I_{3_N} = b\sin\left(\frac{\beta_N b}{2}\right)(\pi m)D_{N_m}$$

$$I_{4_N} = b\cos\left(\frac{\beta_N b}{2}\right)\pi(m-\tfrac{1}{2})D'_{N_m}$$

where

$$D_{N_m} = \frac{(-1)^m}{\left(\frac{\beta_N b}{2}\right)^2-(\pi m)^2} \qquad D'_{N_m} = \frac{(-1)^m}{\left(\frac{\beta_N b}{2}\right)^2-\pi^2(m-\frac{1}{2})^2}$$

The determinantal equation to be solved is

$$H = \begin{vmatrix} U & V \\ X & Y \end{vmatrix} = 0 \quad (29)$$

H is a $4K+2$ square determinant and U, V, X, Y are $K+1$ square sub-determinants. Then

$$U_{NI} = H_{K+1+N,\,K+1+I}$$

$$V_{NI} = H_{K+1+N,\,3K+2+I}$$

$$Y_{NI} = H_{3K+2+N,\,K+1+I}$$

$$Y_{NI} = H_{3K+2+N,\,3K+2+I}$$

Computing procedure

For a fixed βr_1 and trial values of $\omega r_1/c$ the determinant H can be computed. An iterative process is used to predict a more accurate value of $\omega r_1/c$. This process being repeated until the desired accuracy is achieved.

Normalisation of equations

For a large number of space harmonics (K large) the factor $(K_I r_1)^2 = \bar{r}_1^2 - [\beta r_1 + 2\pi I(r_1/p)]^2$ becomes large and negative. Then

$$J_n(K_I r_1) \to I_n(K_I r_1) \to \frac{e^{(K_I r_1)}}{\sqrt{(2\pi K_I r_1)}}$$

becomes extremely large. This causes problems in the determinant evaluation.

To avoid this difficulty the A and B coefficients can be written as

$$\bar{A}_N = A_N J_N(K_N r_1)$$

$$\bar{B}_N = B_N J_N(K_I r_1)$$

In the final eqns. 20 to 29 all the terms of a row are divided by $J_n(K_I r_1)$.

Special case when $L = 0$, no slot modes

In this case the transverse fields of eqns. 5 become

$$\begin{aligned} E_{z_0} &= c_0 S_n^D \cos n\Psi \\ H_{z_0} &= 0 \\ E_{\Psi_0} &= 0 \\ H_{\Psi_0} &= -\mathrm{j} y_0 c_0 \frac{\bar{r}}{(\Gamma_0 r)} S_n^N \cos n\Psi = -\mathrm{j} y_0 c_0 S_n^N \cos n\Psi \end{aligned} \tag{30}$$

Then equating these field components to the field components of region I, eqns. 4, gives the simplified characteristic equation:

$$\frac{S_n^N}{S_n^D}\frac{1}{\bar{r}_1} = \frac{b}{p}\sum_{N=-K}^{K}\left[\frac{J_n'(K_N r_1)}{J_n(K_N r_1)(K_N r_1)} - n^2(\beta_N r_1)^2 \frac{J_n(K_N r_1)}{\bar{r}_1^2 (K_N r_1)^3 J_n'(K_N r_1)}\right]\frac{\sin(\beta_N b/2)}{(\beta_N b/2)} \tag{31}$$

Calculation of the electric and magnetic field components

The matrix equation is

$$\begin{bmatrix} U_{NI} & V_{NI} \\ X_{NI} & Y_{NI} \end{bmatrix} \begin{bmatrix} A_N \\ B_N \end{bmatrix} = 0 \tag{32}$$

To solve this equation we must normalise against one of the coefficients in order to provide an equation of the form

$$\begin{bmatrix} U & V \\ X & Y \end{bmatrix} \begin{bmatrix} A \\ B \end{bmatrix} = \begin{bmatrix} U_{N0} \\ V_{N0} \end{bmatrix} \tag{33}$$

The procedure for solution is best explained with an example. The matrix equation for the $K = 1$ case will be

$$\begin{bmatrix} U_{-1-1} & U_{-10} & U_{-1+1} & V_{-1-1} & V_{-10} & V_{-1+1} \\ U_{0-1} & U_{00} & U_{0+1} & V_{0-1} & V_{00} & V_{0+1} \\ U_{+1-1} & U_{+10} & U_{+1+1} & V_{+1-1} & V_{+10} & V_{+1+1} \\ X_{-1-1} & X_{-10} & X_{-1+1} & Y_{-1-1} & Y_{-10} & Y_{-1+1} \\ X_{0-1} & X_{00} & X_{0+1} & Y_{0-1} & Y_{00} & Y_{0+1} \\ X_{+1-1} & X_{+10} & X_{+1+1} & Y_{+1-1} & Y_{+10} & Y_{+1+1} \end{bmatrix} \begin{bmatrix} A_{-1} \\ A_0 \\ A_{+1} \\ B_{-1} \\ B_0 \\ B_{+1} \end{bmatrix} = 0 \tag{34}$$

The procedure is then:
(i) use either A_0 or B_0 as reference.
(ii) place column $K+1$ (for A_0) or $3K+2$ (for B_0) at the end of the matrix by moving each column on the right of $K+1$ (or $3K+2$) one place to the left.
(iii) Normalise coefficients and ignore the last row, giving (for A_0)

$$\begin{bmatrix} U_{-1-1} & U_{-1+1} & V_{-1-1} & V_{-10} & V_{-1+1} & V_{-10} \\ U_{0-1} & U_{0+1} & V_{0-1} & V_{00} & V_{0+1} & U_{00} \\ U_{+1-1} & U_{+1-1} & V_{+1-1} & V_{+10} & V_{+1+1} & V_{+10} \\ X_{-1-1} & X_{-1+1} & Y_{-1-1} & Y_{-10} & Y_{-1+1} & X_{-10} \\ X_{0-1} & X_{0+1} & Y_{0-1} & Y_{00} & Y_{0+1} & X_{00} \end{bmatrix} \begin{bmatrix} A_{-1}/A_0 \\ A_{+1}/A_0 \\ B_{-1}/A_0 \\ B_0/A_0 \\ B_{+1}/A_0 \end{bmatrix} = 0 \tag{35}$$

for convenience the main matrix has been written in the form of an 'augmented matrix', i.e. the right-hand side of eqn. 33 has been included in the matrix on the left-hand side.
(iv) This matrix equation is solved for A_Ns and B_Ns.

(v) The A_Ns and B_Ns are substituted into eqns. 18 and 19 to obtain d_m and d'_m, then into eqns. 16 and 17 to obtain c_m and c'_m.

$$d_m = \frac{2}{R_n^D} \sum_{I=-K}^{+K} J_n(K_I r_1) \sin\left(\frac{\beta_I b}{2}\right) (\pi m) D_{I_m} B_I \tag{36}$$

$$d'_m = \frac{2}{R_n^{D'}} \sum_{I=-K}^{K} J_n(K_I r_1) \cos\left(\frac{\beta_I b}{2}\right) \pi(m - \tfrac{1}{2}) D'_{I_m} B_I \tag{37}$$

$$\begin{aligned} c_m = \frac{2(\Gamma_m r_1)}{\bar{r}_1 S_n^N} & \left[\sum_{I=-K}^{+K} \left\{ \bar{r}_1 \frac{J'_n(K_I r_1)}{(K_I r_1)} \left(\frac{\beta_I b}{2}\right) \sin\left(\frac{\beta_I b}{2}\right) D_{I_m} A_I \right.\right. \\ & + \left. \frac{(\beta_I r_1)}{(K_I r_1)^2} n J_n(K_I r_1) \left(\frac{\beta_I b}{2}\right) \sin\left(\frac{\beta_I b}{2}\right) D_{I_m} B_I \right\} \\ & \left. - \frac{n}{2} \frac{(\eta_m r_1)}{(\Gamma_m r_1)^2} R_n^D d_m \right] \end{aligned} \tag{38}$$

$$\begin{aligned} c'_m = \frac{2(\Gamma'_m r_1)}{\bar{r}_1 S_n^{N'}} & \left[\sum_{I=-K}^{K} \bar{r}_1 \frac{J'_n(K_I r_1)}{(K_I r_1)} \left(\frac{\beta_I b}{2}\right) \cos\left(\frac{\beta_I b}{2}\right) D'_{I_m} A_I \right. \\ & + \frac{(\beta_I r_1)}{(K_I r_1)^2} n J_n(K_I r_1) \left(\frac{\beta_I b}{2}\right) \cos\left(\frac{\beta_I b}{2}\right) D'_{I_m} B_I \Bigg\} \\ & \left. - \frac{n}{2} \frac{(\eta'_m r_1)}{(\Gamma'_m r_1)^2} R_n^{D'} d'_m \right] \end{aligned} \tag{39}$$

Prediction of radiation characteristics

The far-field radiation patterns can be computed in closed form using the Fourier Transform method. The radiated fields $E_{p\theta}$ and $E_{p\phi}$ are given in terms of the aperture fields $E_r E_\Psi$ by

$$\begin{aligned} E_{p\theta} = \frac{j\frac{\omega}{c} e^{-j\frac{\omega}{c}R}}{2\pi R} & \int_0^{2\pi} \int_0^{r_1} \{E_r \cos(\phi - \Psi) + E_\Psi \sin(\phi - \Psi)\} \\ & e^{j\bar{r} \sin\theta \cos(\phi - \Psi)} r \, dr \, d\Psi \end{aligned} \tag{40}$$

$$\begin{aligned} E_{p\phi} = \frac{j\frac{\omega}{c} e^{-j\frac{\omega}{c}R}}{2\pi R} & \int_0^{r_1} \int_0^{2\pi} \{-E_r \sin(\phi - \Psi) + E_\Psi \cos(\phi - \Psi)\} \\ & \cos\theta \, e^{j\bar{r} \sin\theta \cos(\phi - \Psi)} r \, dr \, d\Psi \end{aligned} \tag{41}$$

Substituting the expressions for the space harmonic field components of the corrugated waveguide given in eqn. 4 in the above equations leads to the required far-field radiation formulae. The resulting equations can be evaluated in closed form by making use of the following integral expressions:

$$\int_0^{2\pi} e^{-jx\cos(\phi-\Psi)} {}^{\cos}_{\sin} n\Psi d\Psi = 2\pi j^n {}^{\cos}_{\sin} n\phi J_n(x) \tag{42}$$

$$\int_0^{r_1} rJ_{n+1}(\bar{r}\sin\theta)J_{n+1}(K_N r)\,dr = \frac{-nJ_n(\bar{r}_1\sin\theta)J_n(K_N r_1)}{K_N\left(\frac{\omega}{c}\right)\sin\theta} + \frac{J_n'(\bar{r}_1\sin\theta)K_N r_1 J_n(K_N r_1) - J_n(\bar{r}_1\sin\theta)J_n'(K_N r_1)\bar{r}_1\sin\theta}{K_N^2 - \left(\frac{\omega}{c}\right)^2\sin^2\theta} \tag{43}$$

$$\int_0^{r_1} rJ_{n-1}(\bar{r}_1\sin\theta)J_{n-1}(K_N r)\,dr = \frac{nJ_n(\bar{r}_1\sin\theta)J_n(K_N r_1)}{K_N\left(\frac{\omega}{c}\right)\sin\theta} + \frac{J_n'(\bar{r}_1\sin\theta)K_N r_1 J_n(K_N r_1) - J_n(\bar{r}_1\sin\theta)J_n'(K_N r_1)\bar{r}_1\sin\theta}{K_N^2 - \left(\frac{\omega}{c}\right)^2\sin^2\theta} \tag{44}$$

Eqns. 43 and 44 are the Lommel integrals.

The resulting closed-form expressions for eqns. 40 and 41 are given for the Nth space harmonic, after some manipulation, by

$$E_{p\theta}^N = \frac{-j^{n+1}\bar{r}_1 e^{-j\frac{\omega}{c}R}\cos n\phi}{(K_N r_1)\left(\frac{R}{r_1}\right)}\left[\bar{A}_N\beta_N r_1 T(\theta) + \frac{\bar{B}_N\bar{r}_1 nJ_n(\bar{r}_1\sin\theta)}{(K_N r_1)\bar{r}_1\sin\theta}\right] \tag{45}$$

$$E_{p\phi}^N = \frac{j^{n+1}\bar{r}_1 e^{-j\frac{\omega}{c}R}\cos\theta\sin n\phi}{(K_N r_1)\left(\frac{R}{r_1}\right)}\left[\frac{\bar{A}_N\beta_N r_1 J_n(\bar{r}_1\sin\theta)n}{(K_N r_1)\bar{r}_1\sin\theta} + \bar{B}_N\bar{r}_1 T(\theta)\right] \tag{46}$$

where

$$T(\theta) = \frac{J_n'(\bar{r}_1\sin\theta)K_N r_1 J_n(K_N r_1) - J_n(\bar{r}_1\sin\theta)J_n'(K_N r_1)\bar{r}_1\sin\theta}{J_n(K_N r_1)\{(K_N r_1)^2 - (\bar{r}\sin\theta)^2\}} \tag{47}$$

The total radiated field is

$$E_\theta(\theta,\phi) = \sum_{-N}^{+N} E_{p\theta}^{N} e^{-j\beta_N z} \cos n\phi \tag{48}$$

$$E_\phi(\theta,\phi) = \sum_{-N}^{+N} E_{p\phi}^{N} e^{-j\beta_N z} \sin n\phi \tag{49}$$

Bibliography

1 ADATIA, N.A., RUDGE, A.W. and PARINI, C.G.: 'Mathematical modelling of the radiation fields from microwave primary-feed antennas', Proc 7th European Microwave conf., Copenhagen, 1977, pp. 329

2 ADATIA, N.A., WATSON, B., DEWEY, R. and RUDGE, A.W.: 'Studies of feeds and reflectors for high gain spacecraft antennas', ERA Report No. RFTC 220676, (ESA Contract 2742/76/NL/AK), Feb. 1978

3 Al-HARIRI, A.M.B., CLARRICOATS, P.J.B. and OLVER, A.D.: 'Low attenuation properties of corrugated rectangular waveguide', *Electron. Lett.*, 1974, **10**, p. 25

4 AL-HARIRI, A.M.B.: 'Low attenuation microwave waveguides', Ph.D. thesis, London University, Oct 1974

5 ASHTON, R.W.: 'Improvements in or relating to corrugated horns (dual depth horn)', UK Patent 1498 905, January 1978

6 AU, H.K.: 'Hybrid modes in conical corrugated horns with narrow flare angle and arbitrary length', *Electron. Lett.*, 1970, **6**, p. 769

7 AUBRY, C. and BITTER, D.: 'Radiation patterns of a corrugated conical horn in terms of Laguerre-Gaussian function', *Electron. Lett.*, 1975, **11**, p. 155–6

8 BACH ANDERSON, J.: 'Propagation in the general, rectangular waveguide', Proc. 5th European Microwave Conf., Hamburg, Sept 1975, pp. 595–599

9 BAHRET, W.F. and PETERS, L.: 'Small aperture – small flare angle corrugated horns', *IEEE Trans.*, 1968, **AP-16**, p. 494

10 BALDWIN, R. and McINNES, P.A.: 'Attenuation in corrugated rectangular waveguide', *Electron. Lett.*, 1971, **7**, pp. 770–772

11 BALDWIN, R. and McINNES, P.A.: 'Corrugated rectangular horns for use as microwave feeds', *Proc. IEE*, 1975, **122**, pp. 465–469

12 BALDWIN, R. and McINNES, P.A.: 'Radiation from rectangular horns with modified boundaries', Proc. European Microwave Conference, 1971, p B7/1

13 BALDWIN, R. and McInnes, P.A.: 'Surface wave radiation from a corrugated horn', *Electron. Lett.*, 1970, **6**, pp. 259–260

14 BALDWIN, R. and McINNES, P.A.: 'A rectangular corrugated feed horn', *IEEE Trans.*, 1975, **AP-23**, pp. 814–817

15 BATHKER, D.A.: 'A stepped mode transducer using homogeneous waveguides', *Trans. IEEE*, 1967, **MTT-15**, pp. 128–130

16 BIELLI, P., DE PADOVA, S. and PAGANA, E.: 'Analysis and synthesis of conical corrugated horns', Proc. 6th European Microwave Conf., Sept 1967, Rome, pp. 153

17 BITTER, D. and AUBRY, C.: 'Sum and difference radiation patterns of a corrugated conical horn by means of Laguerre–Gaussian functions', Proc. 5th European Microwave Conference, 1975, pp. 677–681

18 BRUNSTEIN, S.A.: 'A new wideband feedhorn with equal E- and H-plane beamwidths

and suppressed sidelobes', Jet Propulsion Laboratory. Deep Space Network Space Programs Summary 37–58, July 1969, pp. 61–64

19 BRYANT, G.H.: 'Monopulse multimode feed for military terminals', IEE Conf. on Earth Station Tech. (pub 72), Oct 1970, pp. 245–249.

20 BRYANT, G.H.: 'Propagation in corrugated waveguides', *Proc. IEE,* 1969, **116**, pp. 203–213

21 BUCHMAYER, S.K.: 'Corrugations lock horns with poor beamshapes', *Microwaves*, 1973, **12**, pp. 44–49

22 CALDECOTT, R., MENTZER, C.A. and PETERS, L.: 'The corrugated horn as an antenna range standard', *IEEE Trans.*, 1973, **AP-21**, p. 562

23 CHONG, S.L.: 'Corrugated waveguides for low attenuation transmission', Ph.D. thesis, London University, Now. 1974

24 CHU, T.S. and LEGG, W.E.: 'Gain of corrugated conical horns', *Trans. IEEE*, 1982, **AP-30**, pp. 698–703

25 CLARRICOATS, P.J.B. and ELLIOT, R.D.: 'Multimode corrugated waveguide feed for monopulse radar', *Proc. IEE,* 1981, **128** Pt. H, pp. 102–110

26 CLARRICOATS, P.J.B. and OLVER, A.D.: 'Improvements in the theory and design of corrugated waveguide feeds for low cross polarisation antenna applications', Proc. URSI Int. Symp. on EM Theory, Stanford, June 1977, pp. 233

27 CLARRICOATS, P.J.B. and POULTON, G.T.: 'High efficiency microwave reflector antennas: A review', *Proc. IEEE*, 1977, **65**, pp. 1470–1504

28 CLARRICOATS, P.J.B. and SAHA, P.K.: 'Analysis of spherical hybrid modes in a corrugated conical horn', *Electron. Lett.*, 1969, **5**, pp. 189–190

29 CLARRICOATS, P.J.B. and SAHA, P.K.: 'Propagation and radiation behaviour of corrugated feeds, Pt I Corrugated waveguide feeds', *Proc. IEE,* 1971, **118**, p. 167

30 CLARRICOATS, P.J.B. and SAHA, P.K.: 'Propagation and radiation behaviour of corrugated feeds, Pt II Corrugated conical horn feeds', *Proc. IEE,* 1971 **118**, p. 187

31 CLARRICOATS, P.J.B. and SAHA, P.K.: 'Radiation from wide flare angle scalar horns', *Electron. Lett.*, 1969, **5**, p. 376

32 CLARRICOATS, P.J.B. and SAHA, P.K.: 'Radiation pattern of a lens corrected conical scalar horn', *Electron. Lett.*, 1968, **4**, pp. 592–593

33 CLARRICOATS, P.J.B. and SAHA, P.K.: 'Scalar feeds for earth station antennas', Proc. Conf. on Earth Station Technology, London, Oct 1970, pp. 240–244

34 CLARRICOATS, P.J.B. and SAHA, P.K.: 'Theoretical analysis of cylindrical hybrid modes in a corrugated horn', *Electron. Lett.*, 1969, **5**, pp. 187–189

35 CLARRICOATS, P.J.B. and SENG, L.M.: 'Influence of horn length on the radiation pattern of oblique flare angle corrugated horns', *Electron. Lett.*, 1973, **9**, p. 15

36 CLARRICOATS, P.J.B. and SENG, L.M.: 'Propagation and radiation characteristics of corrugated horns', *Electron. Lett.*, 1973, **9**, pp. 7–9

37 CLARRICOATS, P.J.B., MAHMOUD, S.F. and OLVER, A.D.: 'Cross-polar behaviour of wide-angle corrugated horns', IEEE Symp. on Ant. & Prop., Los Angeles, June 1981, pp. 65–68

38 CLARRICOATS, P.J.B., MAHMOUD, S.F. and OLVER, A.D.: 'Low cross-polar radiation from wide flare angle conical corrugated horns', Proc. 11th European Microwave Conf., Amsterdam, Sept. 1981, pp. 735–739

39 CLARRICOATS, P.J.B., OLVER, A.D. and CHONG, S.L.: 'Attenuation in corrugated circular waveguides, Pt. I Theory', *Proc. IEE*, 1975, **122**, p. 1173

40 CLARRICOATS, P.J.B., OLVER, A.D. and CHONG, S.L.: 'Attenuation in corrugated circular waveguides, Pt. II Experiment', *Proc. IEE*, 1975, **122**, p. 1180

41 CLARRICOATS, P.J.B., OLVER, A.D. and PARINI, C.G.: 'Optimum design of corrugated feeds for low crosspolarisation radiation', Proc. 6th European Microwave Conf., Rome 1976, pp. 148

42 CLARRICOATS, P.J.B., OLVER, A.D. and PARINI, C.G.: 'Radiation method for the measurement of mode conversion levels', *Electron. Lett.*, 1974, **10**, p. 252

43 CLARRICOATS, P.J.B., OLVER, A.D. and SAHA, P.K.: 'Near field radiation characteristics of corrugated horns', *Electron. Lett.*, 1971, **7**, pp. 446–448

44 CLARRICOATS, P.J.B., OLVER, A.D., PARINI, C.G. and POULTON, G.T.: 'Corrugated waveguide feeders for microwave antennas', Proc. 5th European Microwave Conference, Hamburg, Sept. 1975. p. 240

45 CLARRICOATS, P.J.B.: 'Hybrid mode feeds for microwave reflector antennas" in Modern Topics in Electromagnetics and Antennas (Peter Peregrinus, 1979)

46 CLARRICOATS, P.J.B.: 'Similarities in the electromagnetic behaviour of optical waveguides and corrugated feeds', *Electron. Lett.*, 1970, **6**, 6 March,

47 CLARRICOATS, P.J.B.: 'Some recent advances in microwave reflector antennas', *Proc. IEE*, 1979, **126**, pp. 9–25

48 COLEMAN, H.P.: 'Paraboloidal reflector off-set fed with a corrugated conical horn', *Trans. IEEE*, 1975, **AP-23**, pp. 817–819

49 COOPER, P.N.: 'Complex propagation coefficients and the step discontinuity in corrugated cylindrical waveguide', *Electron. Lett.*, 1971, **7**, pp. 135–6

50 DAVIS, D.: 'Corrugations improve monopulse feed horns', *Microwaves*, 1972, pp. 58–63

51 DOANE, J.L.: 'Mode converters for generating the HE_{11} mode from TE_{01} in circular waveguide', *Int. J. Electronics*, 1982

52 DRABOWITCH, S.: 'A new tracking mode coupler using a corrugated feed for satellite communication earth station antenna', Proc. 6th European Microwave Conf., Rome, 1976, p. 287

53 DRAGONE, C.: 'An improved antenna for microwave radio systems consisting of two cylindrical reflectors and a corrugated horn', *Bell Syst. Tech. J.*, 1974, **53**, pp. 1351–1377

54 DRAGONE, C.: 'Characteristics of a broadband microwave corrugated feed: A comparison between theory and experiment', *Bell Syst. Tech. J.*, 1977, **56**, pp. 869–888

55 DRAGONE, C.: 'High frequency behaviour of waveguides with finite surface impedances', *Bell Syst. Tech. J.*, 1981, **60**, pp. 80–116

56 DRAGONE, C.: 'Reflection, transmission and mode conversion in a corrugated feed', *Bell. Syst. Tech. J.*, 1977, **56**, pp. 835–867

57 DYBDAL, R.B., PEAK, W. and PETERS, L.: 'Propagation in corrugated waveguides', *Proc. IEE*, 1970, **117**, pp. 931–932

58 DYBDAL, R.B., PETERS, L. and PEAK, W.: 'Rectangular waveguides with impedance walls', *Trans. IEEE*, 1971, **MTT-19**, p. 2

59 ELLIOT, R.D. and CLARRICOATS, P.J.B.: 'Corrugated waveguide monopulse feed', *Electron. Lett.*, 1980, **16**, pp. 324

60 ELLIOTT, R.S.: 'On the theory of corrugated plane surfaces', *IRE Trans.*, 1974, **AP-2**, pp. 71–81

61 ESTIN, A.J., STUBENRAUCH, C.F., REPJAR, A.G. and NEWELL, A.C.: 'Optimized wavelength-sized scalar horns as antenna radiation standards', *Trans. IEE*, March 1982, **IM-31**, pp. 53

62 FRANK, Z.: 'Very wide band corrugated horns', *Electron. Lett.*, 1975, **11**, 20 pp. 131–133

63 GHOSH, S.. ADATIA, N. and WATSON, B.K.: 'Hybrid mode feed for multiband applications having a dual-depth corrugation boundary', *Electron. Lett.* 1982, **18**, pp. 860–862

64 GHOSH, S.: 'A corrugated waveguide feed for discrete multiband applications having dual depth corrugations', IEEE Int. Symp. on Antennas & Propagation, Quebec, Canada, June 1980, pp. 217–220

65 GRUNNER, R.W.: 'A 4- and 6-GHz prime focus, CP feed with circular pattern symmetry', IEEE Ant. & Prop. Symp. Digest, June 1974, pp. 72–74

66 GUISSARD, A.C. and KANDIL, A.A.: 'A new solution for the fields in corrugated square waveguides', Int. Conf. on Ant. & Prop., Nov. 1978, IEE Conf. Pub. 169, pp. 235–239

67 GUY, R.F.E. and ASHTON, R.W.: 'Cross-polar performance of an elliptical corrugated horn antenna', *Electron. Lett.*, 1979, **15**, p. 400

68 HOCKHAM, G.A. and ELLIOT, R.D.: 'Optimum design of choked waveguide feeds', Int. Conf. on Antennas & Propagation, London, Nov. 1978, IEE Conf. Pub. 169, p. 369

69 HOCKHAM, G.A. and OLVER, A.D.: 'Cross-polarized performance of small corrugated feeds', Proc. IEEE Ant. & Prop. Symp. (Washington), 1978, p. 431

70 HOCKHAM, G.A. and SENG, L.M.: 'Radiation from a parallel plate waveguide with a finite corrugated flange', *Electron. Lett.*, 1975. **11**, p. 292

71 HOCKHAM, G.A.: 'Investigation of a 90 degree corrugated horn', *Electron. Lett.*, 1976, **12**, p. 199

72 HOGG, D.C. *et al.*: 'An antenna for dual-wavelength radiometry at 21 and 32 GHz', *Trans. IEE*, 1979, **AP-27**, pp. 764–771

73 ISSAAC, P.R.: 'Comments on rectangular waveguides with impedance walls', *Trans. IEEE*, 1974, **MTT-22**, p. 972

74 JAMES, G.L. and THOMAS, MacA.B.: 'TE11 to HE11 corrugated cylindrical waveguide mode converters using ring-loaded slots', *Trans. IEEE*, 1982, **MTT-30**, pp. 278

75 JAMES, G.L.: 'Analysis and design of TE11 to HE11 corrugated cylindrical waveguide mode converters using ring-loaded slots', *Trans. IEEE*, 1981, **MTT-29**, pp. 1059–1066

76 JAMES, G.L.: 'Surface reactance of corrugated planes', *Electron. Lett.*, 1979, **15**, p. 751

77 JAMES, G.L.: 'TE11 to HE11 mode converters for small angle corrugated horns', *Trans. IEEE*, 1982, **AP-30**, 1057–1062

78 JANSEN, J.K.M. and JEUKEN, M.E.J.: 'Circularly polarised horn antenna with an asymmetrical pattern', Proc. Colloquium on Microwave Communications, Budapest, 1974, pp. 179–188

79 JANSEN, J.K.M. and JEUKEN, M.E.J.: 'Surface waves in the corrugated conical horn', *Electron. Lett.*, 1972, 8, p. 342

80 JANSEN, J.M.K., JEUKEN M.E.J. and LAMBRECHTSE, C.W.: 'The scalar feed', *Archiv fur Elektronik & Uber*, 1972, **26**, pp. 22–30

81 JANSSEN, M.A. and BEDNARCYK S.M.: 'Pattern measurements of a low-sidelobe horn antenna', *Trans. IEEE*, 1979, **AP-27**, pp. 551–555

82 JEUKEN, M.E.J. and LAMBRECHTSE' C.W.: 'Small corrugated conical horn antenna with wide flare angle', *Electron. Lett.*, 1969, **5**, pp. 489–490

83 JEUKEN, M.E.J. and THURLINGS, L.F.G.: 'The corrugated elliptical horn antenna', IEEE Int. Symp. on Ant. & Prop. Illinois, 1975

84 JEUKEN, M.E.J. and VOKURKA, V.J.: 'Multi-frequency band corrugated conical horn antennas', Proc. European Microwave Conf., Brussels, Sept. 1973

85 JEUKEN, M.E.J.: 'Experimental radiation patterns of the corrugated conical horn antenna with small flare angle', *Electron. Lett.*, **1969, 5**, pp. 484

86 JEUKEN, M.E.J.: 'Frequency independent and symmetric properties of corrugated conical horn antennas with small flare angles', Ph.D. thesis, Eindhoven Univ. of Technology, Netherlands, 1970

87 KAMAL, A.K., GUPTA, S.C. and NAIR, R.A.: 'Radiation characteristics of a dielectric sphere loaded corrugated E-plane sectoral horn', *Trans. IEEE*, 1978, **AP-26**, pp. 625–626

88 KAY, A.F.: 'A wide flare angle horn. A novel feed for low noise broadband and high aperture efficiency antennas', US Air Force Cambridge Research Labs Report 62–757, Oct 1962

89 KAY, A.F.: 'The scalar feed', US Air Force Cambridge Research Labs Report 62–347, AD60169, March 1964

90 KNOP, C.M. and WIESSENFARTH, H.J.: 'On the radiation from an open ended corrugated pipe carrying the HE11 mode', *Trans. IEEE*, 1972, **AP-20**, p. 644

91 KOCH, G.F.: 'Coaxial feeds for high aperture efficiency and low spill-over of paraboloidal reflector antennas', *Trans. IEEE*, 1973, **AP-21**, pp. 164–169

92 LAWRIE, R.E. and PETERS, L.: 'Modifications of horn antennas for low sidelobe levels', *Trans. IEEE*, 1966, **AP-14**, p. 605

93 LOEFER, G.R., NEWTON, J.M., SCHUCHARDT, J.M. and DEES, J.W.: 'Computer analysis speeds corrugated horn design', *Microwaves*, 1976, pp. 53–65

94 LOVE, A.W. (Ed): 'Electromagnetic horn antennas' (IEEE Press, N.Y., 1976)

95 MAHMOUD, S.F. and CLARRICOATS, P.J.B.: 'Radiation from wide flare-angle corrugated conical horns', *Proc. IEE*, 1982, **129**, Pt. H. pp. 221–228

96 MANWARREN, T. and FARRAR, A.: 'Pattern shaping with hybrid mode corrugated horns', *Trans. IEEE*, 1974, **AP-22**, pp. 484–487

97 MARINCIC, A. and MILOVANOVIC, B.: 'Frequency dependence of corrugated horn mode order', Proc. 7th European Microwave Conf., Copenhagen, 1977, p. 306

98 MENTZER, C.A. and PETERS, L.: 'Pattern analysis of corrugated horn antennas', *Trans. IEEE*, 1976, **AP-24**, pp. 304–309

99 MENTZER, C.A.: 'A corrugated horn antenna using V shape corrugations', *Trans. IEEE*, 1975, **AP-23**, p. 93

100 MENTZER, L. and PETERS, L.: 'Properties of cut-off corrugated surfaces for corrugated horn design', *Trans. IEEE*, 1974, **AP-22**, pp. 191–196

101 MINNETT, H.C. and THOMAS, MacA.B.: 'A method of synthesising radiation patterns with axial symmetry', *Trans. IEEE*, 1966, **AP-14**, pp. 654–656

102 MINNETT, J.C. and THOMAS, MacA.B.: 'Propagation and radiation behaviour of corrugated feeds', *Proc. IEE*, 1972, **119**, p. 1280

103 MORRIS, G.: 'A broadband constant beamwidth corrugated rectangular horn', *Trans. IEEE*, 1982, **AP-30**, pp. 966–974

104 MOTO MISUZAW and FUMIO TAKEDA: 'Radiation characteristics of a corrugated conical horn', *Elec. & Comm. in Japan*, 1973, **56B**

105 McINNES, P.A. and BOOKER, D.D.: 'Computer predicted performance of corrugated conical feeds using experimental primary radiation patterns', *Electron. Lett.*, 1970, **6**, pp. 18–20

106 McINNES, P.A.: 'Corrugated conical horns with very wide flare angles', Proc. European Microwave Conf 1971, p. 87

107 NARASIMHAN, M.S. and GOVINDARAJAN, M.S.: 'Eigenvalues of spherical hybrid modes in corrugated conical horns', *Trans. IEEE*, 1974, **AP-22**, pp. 122–123

108 NARASIMHAN, M.S. and RAO, B.V.: 'Diffraction by wide flare angle corrugated conical horn', *Electron Lett.*, 1970, **6**, pp. 469–470

109 NARASIMHAN, M.S. and RAO, B.V.: 'Hybrid modes in corrugated conical horns', *Electron. Lett.*, 1970, **6**, pp. 32–34

110 NARASIMHAN, M.S. and RAO, B.V.: 'Modes in a conical horn: A new approach', *Proc. IEE*, 1971, **118**, pp. 187–192

111 NARASIMHAN, M.S. and RAO, K.S.: 'GTD analysis of the near-field patterns of conical and corrugated horns', *Trans. IEEE*, 1979, **AP-27**, pp. 705–708

112 NARASIMHAN, M.S. and RAO, V.V.: 'Comments on rectangular waveguides with impedance walls', *Trans. IEEE*, 1974, **MTT-22**, p. 973

113 NARASIMHAN, M.S. and RAO, V.V.: 'Radiation characteristics of corrugated E plane sectoral horns', *Trans. IEEE*, 1973, **AP-21**, pp. 320–327

114 NARASIMHAN, M.S. and RAO, V.V.: 'Radiation from wide flare angle corrugated E plane sectoral horns', *Trans. IEEE*, 1974, **AP-22**, pp. 603–608

115 NARASIMHAN, M.S. and SHESHADARI, M.S.: 'Propagation and radiation characteristics of dielectric loaded corrugated dual frequency circular waveguide horn feeds', *Trans. IEEE*, 1979, **AP-27**, p. 858

116 NARASIMHAN, M.S. and SHESHADRI, M.S.: 'GTD analysis of the radiation patterns of wide flare corrugated conical and E-plane sectoral horns', *Trans. IEEE*, 1979, **AP-27**, pp. 276–279

117 NARASIMHAN, M.S.: 'Corrugated conical horns as a space feed for phased array illumination', *Trans. IEEE*, 1974, **AP-22**, p. 720–722

118 NARASIMHAN, M.S.: 'Corrugated conical horns with arbitrary corrugation depth', *Radio & Electrical Engineer*, 1973, **43**, pp. 188–192

119 NARASIMHAN, M.S.: 'Eingenvalues of spherical hybrid modes in corrugated conical horns', *Proc. IEE*, 1973, **120**, pp. 965–967

120 NARASIMHAN, M.S.: 'Radiation from conical horns with large flare angles', *Trans. IEEE*, 1971, **AP-19**, p. 678

121 NARASIMHAN, M.S.: 'Radiation from conical horns', *Int. J. Electronics*, 1969, **27**, p. 377

122 OHTERA, I. and UJILLE, H.: 'Nomographs for phase centers of conical corrugated and TE11 mode horns', *Trans. IEEE*, 1974, **AP-22**, pp. 858–859

123 OLVER, A.D. and DUBROVKA, F.F.: 'Propagation characteristics of open cylindrical corrugated waveguides', Proc. 12th European Microwave Conf., Helsinki, 1982, p. 306

124 OLVER, A.D., CLARRICOATS, P.J.B. and YANG, K.Z.: 'Predicted performance of dual depth corrugated horns', IEEE Int. Symp. on Antennas & Propagation, Houston, Texas, May 1983

125 OLVER, A.D., CLARRICOATS, P.J.B., HOCKHAM, G.A. and PARINI C.G.: 'Design of corrugated feeds for low cross polarisation', Int. Conf. on Ant. & Prop., London, Nov. 1978, IEE Conf. Pub. 169, p.355

126 OLVER A.D.: 'Design of conical corrugated horns', Jet Propulsion Laboratory TDA Progress Report 42–65, 1981, p. 86–93

127 PARINI, C.G. and OLVER, A.D.: 'Accurate prediction of the cross-polar performance of narrow flare angle corrugated horns', Int. URSI Symp. on EM Theory, Munich, Aug. 1980

128 PARINI, C.G., CLARRICOATS, P.J.B. and OLVER, A.D.: 'Cross polar radiation from open ended corrugated waveguides', *Electron. Lett.* 1975, **11**, p. 567

129 PARINI, C.G.: 'Transmission and radiation characteristics of corrugated waveguides', Ph.D. thesis, London University, Oct. 1976.

130 POTTER, P.D.: 'Efficient antenna systems: A new computer program for the design and analysis of high performance conical feedhorns', Jet Prop. Lab Tech. Report 32–1526, Feb. 1973

131 PROFERA, C.E.: 'Improvement of TE11 mode coaxial waveguide and horn radiation patterns by incorporation of a radial aperture reactance', *Trans. IEEE*, 1976, **AP-24**, pp. 203–206

132 RUDGE, A.W., MILNE, K., OLVER, A.D. and KNIGHT, P. (eds): 'The Handbook of Antenna Design', Chapter 4 (Peter Peregrinus, 1983)

133 RULF, B. and HURD, R.A.: 'Radiation from an open-waveguide with reactive walls', *Trans. IEEE*, 1978, **AP-26**, pp. 668–673

134 RUMSEY, V.H.: 'Horn antennas with uniform power patterns around their axis', *Trans. IEEE*, 1966, **AP-14**, pp. 656–658

135 SAKR, L. and SARREMEJEAN, A.: 'The performance of square and conical corrugated horns as monopulse feeds in the millimetric band', Proc. 9th European Microwave Conf., Brighton, 1979, p. 346

136 SCHEFFER, H.: 'Improvements in the development of coaxial feeds for paraboloidal reflector antennas', Proc. 5th European Microwave Conf., Hamburg, 1975, pp. 235–239

137 SHAFAI, L. and HANSEN, J.: 'Matrix formulation of corrugated feeds by using impedance boundary conditions', *Electron. Lett.*, 1977, **13**, p. 310

138 SIMONS, A.J. and KAY, A.F.: 'The scalar feed – A high performance feed for large paraboloid reflectors', IEE Conf. Pub. no21, 1966, pp. 213–217

139 SPIKE, W.: 'Realisation of a hybrid mode secondary focus feed for the 100 m radiotelescope at Effelsburg', Proc. 1971 European Microwave Conf. pB6/1

140 TAKEDA, F. and HASHIMOTO, T.: 'Broadbanding of corrugated conical horns by means of the ring-loaded corrugated waveguide structure', *Trans. IEEE*, 1978, **AP-26**, pp. 367–372

141 TAKEICHI, Y., HASHIMOTO, T. and TAKEDA, F.: 'The ring-loaded corrugated waveguide', *Trans. IEEE*, 1971, **MTT-19**, pp. 947–950

142 TERZUOLI, A.J.: 'VSWR properties of E-plane dihedral corrugated horns', *Trans. IEEE*, 1978, **AP-26**, pp. 239–243

143 TERZUOZLI, A.J., RICHMOND, J.H. and PETERS, L.: 'Shielding effectiveness of corrugations in corrugated horns', *Trans. IEEE*, 1978, **AP-26**, p. 654

144 THOMAS, MacA.B. and GREENE, J.: 'A curved-aperture corrugated horn having very low cross-polar performance', *Trans. IEEE*, 1982, **AP-30**, pp. 1068–1072

145 THOMAS, MacA.B.: 'Bandwidth properties of corrugated conical horns', *Electron. Lett.*, 1969, **5**, pp. 561–563

146 THOMAS, MacA.B. and GREENE, K.J.: 'Minimizing the cross-polar energy from corrugated horns', *IREE Conv.*, 1979, pp. 13–16

147 THOMAS, MacA.B. and MINNETT, H.C.: 'Modes of propagation in cylindrical waveguides with anisotropic walls', *Proc. IEE*, 1978, **125**, pp. 929–932

148 THOMAS, MacA.B. and MINNETT, H.C.: 'Propagation in cylindrical waveguides with anisotropic walls', Radiophysics Publication RPP 1346, CSIRO, Sydney, Jan. 1977

149 THOMAS, MacA.B.: 'Design of corrugated conical horns', *Trans. IEEE*, 1978, **AP-26**, pp. 367–372

150 THOMAS, MacA.B.: 'Mode conversion using circumferentially corrugated cylindrical waveguide', *Electron. Lett.*, 1972, **8**, p. 394

151 THOMAS, MacA.B.: 'Prime focus focus one and two hybrid mode feeds', *Electron. Lett.*, 1970, **6**, pp. 460–461

152 THOMAS, MacA.B.: 'Theoretical performance of prime focus paraboloids using cylindrical hybrid mode feeds', *Proc. IEE*, 1971, **118**, pp. 1539–1549

153 THOMAS, R.F. and BATHKER, D.A.: 'A dual hybrid mode feedhorn for DSN antenna performance enhancement', Jet Prop. Lab. Deep Space Network Progress Report 42–22, 1974, pp. 101–108

154 VIGGH, M.: 'Study of design procedures and limitations for monopulse scalar feeds', Rome Air Development Centre Report No. RADC-Tr-69-303, ASTIA Document 862515, Nov. 1969

155 VU, T.B. and HIEN, N.V.: 'A new type of high performance monopulse feed', *Trans. IEEE*, 1973, **AP-21**, pp. 855–857

156 VU, T.B. and VU, Q.M.: 'Corrugated horns – recent developments', Proc. 1971 European Microwave Conf. pB6/3

157 VU, T.B. and VU, Q.M.: 'Optimum feed for large radiotelescopes: Experimental results', *Electron. Lett.*, 1970, **6**, pp. 159–160

158 VU, T.B.: 'Corrugated horns as high performance monopulse feed', *Int. J. Electronics*, 1973, **32**, pp. 433–444

159 VU, T.B.: 'On the bandwidth characteristics of corrugated feed horns', *Int. J. Electronics*, 1970, **29**, p. 449

160 VU, T.B.: 'Optimisation of performance of corrugated feed for paraboloidal antenna', *Int. J. Electronics*, 1971, **30**, p. 189

161 WATSON, B.K., DANG, N.D. and GHOSH, S.: 'A mode extraction network for RF sensing in satellite reflector antennas', Int. Conf. on Ant. & Prop. IEE Conf. Publ. 195, April 1981, pp. 323–327

162 WATSON, B.K., DANG, R., RUDGE, A.W. and OLVER, A.D.: 'Compact low cross-polar corrugated feed for ECS', IEEE Int. Symp. on Ant. & Prop., Quebec, June 1980

163 WILLIAMS, W.F. and WITHINGTON, J.R.: 'A common aperture S- and X-band feed for the deep space network', Antenna Applications Symp. University of Illinois, Illinois, Sept 1979

164 WOHLLEBEN, R., MATTES, H. and LOCHNER, O.: 'Simple small primary feed for large opening angles and high aperture efficiency', *Electron. Lett.*, 1972, **8**, pp. 474–476

165 WOHLLEBEN, R., WIELHEBINSKI, R. and MATTES, H.: 'Feeds for the 100 m Effelsberg telescope', Proc. European Microwave Conf., 1971, pp. B5/5
166 YOSHIRO, T., TSUTOMV, M. and FUMIO, T.: 'The ring loaded corrugated waveguide', *Trans. IEEE*, 1971, **MTT-19**, pp. 947–950
167 ZACHARIAH, E.J., VASUDEVAN, K. and NAIR, K.G.: 'Metal flanges with more parameters for beam shaping', *Trans. IEEE*, 1979, **AP-27**, pp. 708–711
168 FASOLD, D. and GALL, H.: 'Theory and computer program for the radiation pattern prediction of conical corrugated horns', *Nachrichtentechnische Zeitschrift,* 1978, **31**, pp. 294
169 FASOLD, D. and PECHER, H.: 'Gain of rectangular horns', *Microwave Journal,* March 1979, pp. 76–79
170 FASOLD, D. and PECHER, H.: 'Berechnung und Auslegung von Rechteckrillenhornstrahlern', *Mikrowellen Magazin,* 1981, pp. 39
171 KUHN, E. and HOMBACH, V.: 'Computer aided analysis of corrugated horns with axial or ring-loaded radial slots', 3rd Int. Conf. on Ant. and Prop., IEE Conf. Pub. 219, April 1983, pp. 879–884
172 JAMES, G.L.: 'Propagation and radiation properties of corrugated cylindrical coaxial waveguides', *Trans. IEEE,* 1983, **AP-31**, pp. 477–483
173 SHAFAI, L., KISHK, A.A., ITTIPIBOON, A. and BRIDGRES, E.: 'Performance of corrugated feeds with simple corrugation shapes', IEEE Int. Symp. on Ant. and Prop., Houston, Texas, May 1983, pp. 586–588
174 OLVER, A.D., YANG, K.Z. and CLARRICOATS, P.J.B.: 'Dual depth corrugated horn design', Proc. 13th European Microwave Conf., Nuremburg, Sept 1983, pp. 879–884
175 SRIDHAR, N. and SRIVASTAVA, G.P.: 'Large bandwidth realisation in dual mode corrugated guide feeds', Proc. 13th European Microwave Conf., Nuremburg, Sept. 1983, pp. 891
176 WANG, H.L. and ZHANG, R.Z.: 'Complex eigenvalues of corrugated waveguide', Proc. 13th European Microwave Conf., Nuremburg, Sept. 1983, pp. 885–890
177 MAHMOUD, S.F.: 'Mode conversion in profiled corrugated conical horns', *Proc. IEE,* Pt H, Oct 1983, 130

GENERAL REFERENCES

178 ABRAMOWITZ, M. and STEGUN, I.A.: 'Handbook of Mathematical Functions' (Dover Publications, New York, 1965)
179 BARTLETT, H.E. and MOSELEY, R.E.: 'Dielguides – high efficiency low noise antenna feeds', *Microwave J.*, 1966, **9**, pp. 53–58
180 BARTLETT, H.E., MOSELEY, R.E. and PIETSCH: United States Patent Nos. 3414903 (Dec. 68), 3430244 (Feb. 1969) and 3611391 (Oct. 1971)
181 BEM, D.J.: 'Electric field distribution in the focal region of an offset paraboloid', *Proc. IEE*, 1969, **116**, pp. 579–584
182 CLARRICOATS, P.J.B. and SALEMA, C.E.R.C.: 'Antennas employing conical-dielectric horns: Part I – Propagation and Radiation characteristics of dielectric cones, Part II – The Cassegrain antenna', *Proc. IEE,* 1973, **121**, pp. 741–756
183 HARRINGTON, R.F.: 'Time harmonic electromagnetic field' (McGraw Hill, New York, 1961)
184 IEEE Standard Test Procedures for Antennas, IEEE Standard 149–1979, IEEE, New York, 1979
185 LUDWIG, A.C.: 'The definition of cross-polarisation', *IEEE Trans.*, 1973, **AP-21**, pp. 116–119
186 MINNETT, H.C. and THOMAS, MacA.B.: 'Fields in the image space of symmetrical focussing reflectors', *Proc. IEE*, 1968, **115**, p. 1419

187 PHILLIPS, C.J.E. and CLARRICOATS, P.J.B.: 'Optimum design of a Gregorian corrected spherical reflector antenna', *Proc. IEE*, 1970, **117**, p. 718

188 POTTER, P.D.: 'A new horn with suppressed sidelobes and equal beamwidths', *Microwave J.* 1963, **6**, pp. 71–76

189 RUDGE, A.W. and ADATIA, N.A.: 'A new class of primary-feed antennas for use with off-set parabolic reflector antennas', *Electron. Lett.*, 1975, **11**, pp. 597–599

190 RUDGE, A.W. and ADATIA, N.A.: 'Matched-feeds for offset parabolic reflector antennas', Proc. European Microwave Conf. 1976, p. 143

191 RUDGE, A.W., MILNE, K., OLVER, A.D. and KNIGHT, P.K. (eds): 'The Handbook of Antenna Design', Vol. I (Peter Peregrinus, 1982)

192 THOMAS, MacA.B., MINNETT, H.C. and VU, T.B.: 'Fields in the focal region of a spherical reflector', *IEEE Trans.*, 1969, **AP-17**, pp. 229–231

Program 1

```
1000 REM***Program to compute the propagation
1010 REM     and attenuation characteristics of
1020 REM     corrugated waveguide using surface
1030 REM     impedance formulation of propagation
1040 REM     coefficient and pertubation formula
1050 REM     for attenuation.
1060 DIM C(20), F(20)
1070 REM
1080 PI=3·14159
1090 REM Input Data
1100 INPUT N,RA,R1,D,TT,RH
1110 INPUT FR,C(1)
1120 IF FR=-1 GOTO 1090
1130 R=1/(RA/R1+1)
1140 WW=FR*2*PI*R1/0·3
1150 IF C(1)=-1 THEN C(1)=SQR (WW*2-5·7)-0·01
1160 REM
1170 REM Print Titles
1180 PRINT:PRINT
1190 PRINT SPC(16);"Corrugated Waveguide"
1200 PRINT:PRINT" Azimuthal No. =";N
1210 PRINT" R1/R0          =";R
1220 PRINT" R1             =";R1 "metres"
1230 PRINT" Slot depth     =";RA;"metres"
1240 PRINT" Slot Width     =";D;"metres"
1250 PRINT" Ridge Width    =";TT;"metres"
1260 PRINT" Conductivity   =";RH
1270 PRINT
1280 CO=0: T=0: M=0: V=1: CH=1
1290 AC=0·001
1300 GOTO 1310
1310 REM
1320 REM Function Evaluation
1330 BA=C (M+1)
1340 K2=WW*WW-BA*BA
1350 SS=SGN(K2)
1360 IF SS=1 AND ABS (K2) >0·01 THEN K=SQR (K2)
1370 IF SS=-1 AND ABS (K2) >0·01 THEN K=SQR (-K2)
1380 IF SS=0 OR ABS (K2) <=0·01 GOTO 2040
1390 X=WW/R: MM=1: GOSUB 2220
1400 Y1=J1: Y2=J2: Y3=J3: Y4=J4
1410 X=K: MM=SS: GOSUB 2220
1420 Z1=J1:Z2=J2:Z3=J3:Z4=J4
```

```
1430 X=WW: MM=1: GOSUB 2220
1440 SN=J2*Y3-Y1*J4
1450 SD=J1*Y3-Y1*J3
1460 S=WW*SN/SD
1470 FT=K*Z2/Z1
1480 M=M+1
1490 F(M)=K*K2*SN*Z1*Z2/WW-K*K*Z2*Z2*SD
1500 F(M)=F(M)+N*N*BA*BA*Z1*Z1*SD/(WW*WW)
1520 IF V=1 GOTO 1550
1530 IF M>=2 GOTO 1660
1540 GOTO 1310
1550 REM
1560 REM Track to find root
1570 BA=BA+0·02*CH
1580 IF BA>=WW THEN BA=BA-0·01: CH=-1
1590 IF M=1 GOTO 1340
1600 SS=SGN(F(M): ST=SGN(F(M-1))
1610 IF SS=ST THEN F(M-1)=F(M): M=M-1: GOTO 1340
1620 C(1)=BA-0·005*CH
1630 C(2)=BA-0·01*CH
1640 V=0: T=T+1: M=0
1650 GOTO 1310
1660 REM
1670 REM Lagrangian inverse interpolation
1680 H=C(M): IF M>=20 GOTO 2040
1690 FOR I=1 TO M-1
1700 C(M)=(C(I)*F(M)-H*F(I))/(F(M)-F(I))
1710 H=C(M): NEXT
1720 IF H<0 THEN M=M-1: GOTO 2040
1730 C(M+1)=H
1740 IF ABS (C(M-1)-C(M))/ABS(C(M))>AC GOTO 1310
1750 CO=0: BA=H
1760 REM
1770 REM Attenuation
1780 RS=SQR(0·04*FR/RH)*PI
1790 P1=2/(PI*PI*SD*SD)-(S+1)*(S+1)/2
1800 P1=P1-(WW*WW-1-N*N)/2
1810 A1=2*PI*RS*Z1*Z1*R1*R1/(WW*WW*PI*PI*14400)
1820 P1=A1*P1
1830 P2=A1*2*D*R/(PI*PI*R1*SD*SD)
1840 DE=-BA*N/(WW*FT)
1850 P3=N*BA*DE/WW+FT
1860 P3=DE*DE+P3*P3/K2
1870 P3=A1*WW*WW*P3*TT/(2*R1)
```

```
1880 H=(D+TT)*BA/(2*R1*SIN(BA*D/(2*R1)))
1890 H=H*H
1900 P1=H*P1: P2=H*P2
1910 PO=BA*(1+DE*DE)*(FT+K2/2-N*N/2+FT*FT/2)/WW
1920 PO=PO+N*DE*(1+BA*BA/(WW*WW))
1930 PO=PO*PI*WW*WW*Z1*Z1*R1*R1*/(K2*K2*120*PI)
1940 AL=4·343*(P1+P2+P3)/(PO*(D+TT))
1950 REM
1960 REM Print
1970 PRINT" Frequency      =";FR;"GHz"
1980 PRINT" k*r1           =";WW
1990 PRINT" Beta*r1        =";BA
2000 PRINT" Hybrid factor  =";DE
2010 PRINT" Attenuation    =";AL;"dB/m"
2020 PRINT
2030 GOTO 1110
2040 REM
2050 REM Modification
2060 CO=CO+1
2070 IF CO>2 GOTO 2110
2080 PRINT" MODIFICATION"
2090 BA=BA-0·002
2100 GOTO 1340
2110 REM
2120 REM Error
2130 PRINT" ERROR"
2140 GOTO 1090
2150 END
2160 REM Subroutine BESSEL (X,N,M,J)
2170 REM Computes J(X),dJ(X)/dX,Y(X),dY(X)/dX,
2180 REM Order N, when M=+1. IF M=-1 the
2190 REM corresponding modified functions I & K
2200 REM are computed. Values are stored in J1,J2
2210 REM J3,J4 in the above order.
2220 J3=0: J5=0: J6=0: J2=1: J7=-1
2230 FORI=2*INT(0·8*X+4) TO1STEP-1
2240 J1=2*I*J2/X-J3*MM: J3=J2: J2=J1
2250 IF (-1)^I<>1 GOTO 2270
2260 J7=-J7*MM: J5=J7*J3/I+J5: J6=J6+J3: GOTO 2280
2270 IFMM=-1 THEN J6=J6+J3
2280 NEXTI
2290 J6=J6+J6+J1: J5=J5*J7: J7=1/J6
2300 IFMM=-1THENJ7=EXP(X)*J7
2310 J8=-0·7267604*MM+3·2732395
```

```
2320 J6=LOG(X)-0·1159315: J6=0·25*J1*J6+J5*MM
2330 J5=J6*J8*MM
2340 J1=J2*J7: J2=J3*J7: J3=J5*J7
2350 IFX*MM>=-1GOTO 2440
2360 J5=0·0091894
2370 DATA-0·668098, 0·2184518,-0·4262633, 0·5575368
2380 DATA-0·5247277, 0·379241,-0·229985, 0·1344596
2390 DATA-0·0913909, 0·0881113,-0·1566642, 1·2531414
2400 RESTORE 2370: FORI=1TO12
2410 READJ6: J5=J5/X+J6
2420 NEXTI
2430 J3=EXP(-0·5*LOG(X)-X)*J5
2440 J4=J3*J2-0·25*J8/X: J4=-J4*MM/J1: J2=-MM*J2
2450 IFN=OGOTO2510
2460 FORI=1TON
2470 J5=J1: J8=J3
2480 J1=((I-1)*J5/X-J2)*MM: J3=(I-1)*J8/X-J4
2490 J2=J5-I*J1/X: J4=J8*MM-I*J3/X
2500 NEXTI
2510 RETURN
```

Program 2

```
1000 REM***Program to compute co-polar and
1010 REM    cross-polar patterns from open-
1020 REM    ended corrugated waveguide.
1030 REM    Valid for kr1 large.
1040 PRINT
1050 PRINT" Radiation Pattern of Corrugated Waveguide"
1060 PRINT
1070 INPUT" Inner radius in Wavelengths =";R1
1080 KR=6·2832*R1
1090 INPUT"Slot depth in Wavelengths =";A
1100 PRINT
1110 KA=6·2832*A
1120 Y=1/TAN(KA)
1130 PRINT" Slot admittance =";Y
1140 U0=2·405
1150 U=U0*(1-Y/(2*KR))
1160 U2=U0*U0
1170 G=1-U2*(Y/(2*KR)-Y*Y*U2*(4+U2)/(8*KR*KR))
1180 PRINT" Hybrid factor =";G: PRINT
1190 INPUT" Increment angle, Final angle =";DT,FT
1200 X=U: GOSUB 1400
1210 U0=J0: U1=J1
```

```
1220 PRINT:PRINT"   ANGLE    CO-POLAR    X-POLAR"
1230 PRINT SPC(15);"(dB)";SPC(11);"(dB)"
1240 FOR T=0 TO FT STEP DT
1250 V=KR*SIN(T/57·295): X=V
1260 IF T=0 THEN Z=0: J0=1: J1=0: GOTO 1280
1270 GOSUB 1400
1280 N2=-A*A/(V*V-U*U)
1290 NO=(V*U0*J1-U*J0*U1)*N2
1300 IF T=0 GOTO 1340
1310 N2=(V*(-U0+2*U1/U)*J1-U*(-J0+2*J1/V)*U1)*N2
1320 Z=(G-1)/(G+1)*N2
1330 XP=8·686*LOG(ABS(Z))
1340 CP=8·686*LOG(ABS(NO+Z))
1350 IF T=0 THEN CO=CP: XP=-100+CO
1360 CP=CP-CO: XP=XP-CO
1370 PRINT T,CP, XP
1380 NEXT
1390 GOTO 1040
1400 REM Subroutine to compute Bessel functions
1410 REM J0=JO(X) and J1=J1(X)
1420 REM
1430 J2=0; J3=0; J1=1
1440 FOR I=2*INT(0·8*X+4) TO 1 STEP -1
1450 J0=2*I*J1/X-J2: J2=J1: J1=J0
1460 IF (-1)^I=1 THEN J3=J3+J2
1470 NEXTI
1480 J3=J3+J3+J0: J4=1/J3
1490 J0=J0*J4: J1=J2*J4
1500 RETURN
```

Index

Airy, George 11
Anechoic chambers 171–175
  Alignment with laser 174
  Assessment of performance 175–177
  Funnel 173
  Instrumentation 175
  Rectangular 173
Aperture diameter (choice of) 142
Attenuation in c. waveguides 2, 20, 40–44, 143
  Attenuation coefficient 41–44
  Balanced-hybrid condition 43–44
  Power flow 41
  Power loss 42
  Poynting vector 41

Balanced hybrid condition 16, 27–30, 62, 90, 105, 106, 157–158
Balanced hybrid frequency 115–116, 119, 122, 167
Beam waist of Gaussian beam 95–96
Beam waveguide feed system 97, 102
Beamwidth design curves 107
Bessel function computation 159–160
Bibliography 213–221
Blockage loss 139
Broadband horns 98, 153–154

Cassegrain antenna feeds 5, 20
Cassegrain reflector antennas 1, 97, 138, 141
Casting of horns 165
Characteristic equations for propagation
  Conical c. horns 61–62
  Cylindrical c. waveguides 25, 27–29, 202–208
  Space harmonic 202–208
  Surface impedance 25, 27–29
Choked feeds 4
Choked flange 145
Compact Antenna Range 172
Compact horns 49, 97–98, 127–128, 155–157
Computer programs 222–226
Conical c. waveguides 58–96, 158–159
  (see also Wide flare angle horns)
  Bandwidth 68
  Computer prediction 158–159
  Frequency dependence of crosspolar radiation 68
  Gain 68–69
  Higher order mode generation 69–86
  Hybrid factor 61–62
  Modal characteristics 58–65
  Modal equation 61–63
  Modal fields 60–61, 64–65
  Propagation 58–65
  Radiation by Kirchoff-Huygen method 86–89
  Radiation by Laguerre-Gaussian method 94–96
  Radiation by SWEX method 65–69
  Radiation by waveguide approximation 90–94
Copolar pattern measurement 177–180
Copolar pattern with simple formula 167
Copolar radiation characteristics 100–114
  Beamwidth design curves 107–109
  Definition 101
  Frequency behaviour 107
  Gaussian pattern 101–103
  Higher order modes 111–114
  Near-field patterns 109–111
  Normalised patterns 105
  Optimum edge taper 102
Corrugated horn design 141–156
Corrugations
  Geometry (choice of) 142
  Influence on crosspolarisation 114–122

  Orientation 121–122
  Shape 168–169
  Width 120
Crosspolar pattern measurement 177–180
Crosspolarisation
  Bandwidth 115, 121–122
  Conical c. horns 64–65, 68–69, 78, 94
  Cylindrical c. waveguides 45–50
  Cylindrical waveguide approximation 90–94
  Definition 101
  Experimental results 120
  Near-field patterns 122–123
  Parametric study 117–122
  Radiation characteristics 114–122
  Sources of crosspolar power 108
  Space harmonic model results 114–122
  Surface impedance model 114, 120
Cylindrical c. waveguides 20–57, 143, 160–161
  Computer prediction 157–158

Design 97–162
Design guidelines 142–143
Dielectric cone feed 1, 5
Dielectric waveguide feeds
  Hybrid mode feed 9
  Rectangular cross-section 10
Dielguide 1
Directivity
  Antenna 135
  Circular aperture 135
  Corrugated horns 135–137
  Maximum 136
  Peak 135
Dual depth horns 2, 122, 154–155
Dual hybrid mode horn 153
Dual-shaped reflector 16, 138

Earth-station antennas 5, 16
Edge taper 102, 138
Efficiency
  Blockage 8
  Crosspolarisation 8, 138
  Earth stations 5
  Focal fields 13
  Higher order mode improvement 55
  Illumination 8, 137–138, 140
  Maximum 16–17
  Microwave-relays 5
  Noise-temperature 5
  Open-ended c. waveguides 138
  Optimum edge taper 102
  Phase 8, 138
  Sidelobe levels 5
  Spillover 5, 8, 138, 140
  Subreflector shape 7
  Surface error 8
  Wide flare angle horns 138, 140
Electrical design parameters 99
Electroforming 164–165
Elliptical c. horns 2, 10, 193–198
  Attenuation 193
  Characteristic equation 197
  Crosspolar peformance 193
  Field components 195–196
  History 193
  Mathieu equations 194–195
  Mathieu functions 195–198
  Propagation characteristics 198
  Radiation characteristics 198
  Wave equation 194

Far-field distance 109, 172
Far-field test range 171
Fields in a cylindrical waveguide 36–39
Five horn cluster 150
Flare angle (choice of) 142–143
Flare section 73–86, 99, 122–128
Floquet's theorem 200
Focal fields of a reflector 10–19
  Efficiency 14
Four horn cluster 150
Fourier Transform method 8, 21, 37, 43–55, 109, 143, 160, 190–191, 210–212

G/T parameter 140–141
GTD 22
Gain
  Antenna 5, 135
  Corrugated horns 135–137
  Measurement 180
  Reflector antenna 137
  Standard (use as) 97, 136
Gaussian beam 94–96, 102–103
Gaussian pattern 101
Gregorian subreflector 17, 97

Higher order modes
  Attenuation 44
  Conical horn radiation 68–69
  Copolar patterns 111–114
  Crosspolar patterns 114, 122
  Generation 69–86

Generation with constant flare angle 73–78, 127
Generation with variable flare angle 78–86, 128, 157
Improved efficiency with 54, 138, 151–153
Matched feeds 55, 153
Multimode waveguides and horns 54–57, 150–153
Propagation characteristics 23–25, 62–63
Throat region 73, 135, 146
History of c. horns 1–3
Hybrid factor 25–32, 61, 116
Hybrid-mode feeds 5–19
Balanced-hybrid condition 17
Bandwidth 9
Corrugated feeds 5, 9
Dielectric waveguides 9
Dielectric-cone feed 5
Impedance of boundary 7

INTELSAT-A earth stations 97, 102
Ideal feed 7, 137
Illumination loss 5
Illumination of a reflector 5–10

Junctions in c. waveguides 50–53, 128–135

Kirchhoff-Huygen method 22, 43, 86–89, 96
Comparison with SWEX method 87–89

Laguerre-Gaussian method 58, 94–96
Legendre function approximation 62
Low noise systems 138
Ludwig crosspolarisation definition 46, 101

Machining of horns 163–164
Manufacture of c. horns 163–169
Casting 165
Electroforming 164–165
Machining 163–164
Shape of corrugations 168–169
Spark erosion 165
Thermal contraction 166
Tolerances 163, 167–168
Matched feeds 2, 55, 111, 153
Measurement of horns
Copolar patterns 177–180
Crosspolar patterns 177–180
Gain 180
Phase patterns 180
Measurements 169–180

Microwave-relay applications 5
Modal expansion method 50
Mode conversion 114, 73–86, 146
Multifrequency horns 97–98, 154–155
Multimode horns and waveguides 13, 16, 18, 54–57, 98, 150–153
Efficiency improvement 54
Field equations 55–57
Tracking capability 54

NASA Deep Space Network antennas 154
Narrow flare angle horns 49, 94, 97–98, 141–143
(see also Cylindrical c. waveguides)
(see also Radiation from c. waveguides)
Attenuation 143
Copolar patterns 103–114
Crosspolar patterns 114–120
Design procedure 142–143
Numerical prediction 160–162
Near-field patterns 89, 109–111, 122–123
Near-field test range 172
Noise temperature 5, 140–141
Numerical prediction 157–162
Bessel functions 159–160
Computer programs 159, 162
Conical c. waveguides 158–159
Corrugated circular waveguide 157–158
Propagation characteristics 157–160
Radiation characteristics 160–162

OTS satellite 143

Paraboloidal reflector, offset 16, 55
Paraboloidal reflector, prime focus 151
Parkes radiotelescope 55
Pattern Comparison method 176–177
Phase centre 109, 124–126, 146
Near-field 124, 126
Profile horn 125
Wide flare horn 125
Phase pattern measurement 180
Potter horn 4, 54
Prime focus feeds 20
Profile horns, 20, 49, 78–86, 97–98, 125–128, 155–157
Mode conversion along 78–86
Phase centre 125, 157
Propagation in conical c. waveguides 58–65
Modal equations 61–63
Modal fields 60–61, 64–65
Propagation in cylindrical c. waveguide 22–34, 157–160, 202–208

Balanced-hybrid conditions 27–30
Cut-off condition 25–26
Fast-wave to slow-wave transition 30
High-frequency cut-off 30–32
Numerical prediction 157–160
Short-circuit condition 30
Space harmonic analysis 202–208
Space harmonics (effect of) 32–34
Surface impedance approach 24
Pyramidal horn 174

Quiet zone 174

Radar absorbing material 174
Radar antennas 10, 17
Ultra-low sidelobe 17
Radiation from c. waveguides 44–50, 160–162, 210–212
Flange (effect of) 44
Fourier Transform method 44–50, 210–212
GTD 44
Kirchhoff-Huygen method 44
Numerical prediction 160–162
Radiation formulae 47
Radio astronomy antennas 1, 16, 97, 138
Rayleigh, Lord 14
Rectangular c. horns 2, 10, 181–193
Attenuation 182
Balanced hybrid condition 185, 187, 188, 191
Beamwidth 191
Crosspolarisation 192
Cut-off frequencies 188
E plane sidelobes 181
Field components 184
Four-wall horn 181–182, 188–190
High frequency cut-off 188
History 2, 181–182
Horn flare (effect of) 192
Magnetic wall 185
Manufacture 181
Mode nomenclature 182
Propagation characteristics 187–188, 190
Radiation characteristics 190–193
Square horn 190–191
Surface impedance 185
Throat region 181, 182, 187, 192
Two-walled horn 181–188
VSWR characteristics 193
Reflectivity level 172, 176
Ring-loaded slots 133–135

Satellite communications 2
Satellite-broadcast antennas 10
Satellite-communication earth station 5
Scalar feeds 97
Separation of variables assumption 62
Shaped beam antennas 10
Shaped pattern horns 151–153
Sidelobe levels 5, 141
Slots
Geometry (choice of) 142
Influence on crosspolarisation 114–122
Orientation 121–122
Shape 168–169
Wide for millimeter wavelengths 120
Small aperture horns 143–145
Space attenuation loss 151
Space harmonic analysis 199–212
Calculation of field components 209–210
Computing procedure 208
Field components 200–202
Propagation equation 202–208
Radiation characteristics 210–212
Space-harmonics 20, 32–34, 114, 199–212
Field in cylindrical waveguide 39
Spacecraft antennas 17, 57, 97
Spark erosion technique 165
Spherical reflector 17–19
Spherical wave expansion method (SWEX) 58–86, 107, 109, 135
Spillover 138
Spillover loss 5
Spillover noise 140–141
Standard gain corrugated horn 136
Subreflector shape 7
Surface error loss 139
Surface impedance representation 20, 24

Test ranges 170–177
Anechoic chamber design 172–175
Assessment of performance 175–177
Types 171–172
Testing of corrugated horns 169–180
Thermal contraction 166
Throat region of horn 73, 99, 128–135, 143, 146–148, 154
Design procedure 128–135
Frequency band 133
Mode conversion 133
Mode generation 73
Non-linear profile 132
Reflection coefficient 129
Slot geometry 131–133

Tracking modes 135
Tolerances 163, 167–168
Tracking horns 150–151

VSWR method of test range assessment 176

Waveguide approximation for horns 58, 90–94, 161
Spherical phase factor 92

Wide flare angle horns
(see also Conical c. waveguides)
Copolar patterns 103–111
Corrugation geometry 146
Crosspolar patterns 114, 120–121
Efficiency 146
Mode conversion 77
Numerical prediction 158–159, 161–162
Phase centre 125, 146
Throat region 146–148

www.ingramcontent.com/pod-product-compliance
Lightning Source LLC
LaVergne TN
LVHW011758200926
848484LV00003B/732